Applied

Rock Mechanics

and

Ground Stability

a textbook for engineering students and practitioners

Second Edition

by

D.H. Steve Zou, Ph.D., P. Eng.

Professor of Mineral Resource Engineering
Dalhousie University
Halifax, Nova Scotia
Canada

Canamaple Academia Services
Camdemia   Halifax, Canada

This textbook is intended for use by engineering students, who have adequate knowledge in engineering geology and mining engineering. It may also be used as a reference book by practical engineers and other professionals working in rock engineering.

ISBN: 978-0-9948791-5-8 (print)
ISBN PDF: 978-0-9948791-6-5 (ebook)

Applied Rock Mechanics and Ground Stability, **Second Edition**
D.H. Steve Zou, Ph.D., P. Eng.

Revised in May 2020.

Subject headings: rock mechanics / mining / geotechnical engineering

Collection in Library and Archives Canada

Published and distributed by
Canamaple Academia Services
Halifax, Canada, B3M 2Y2
http://press.camdemia.ca, press@camdemia.ca

Disclaimer:

While all care is taken to ensure the accuracy and correctness of the information contained in this textbook, the author and the publisher shall not be liable for any damage to property or persons as a result of using the information contained herein.

To my parents

Acknowledgement

I must admit that there were a lot more work in writing a textbook than I ever imagined. What I did not realize was that there always seemed to be something more important than completing a book in one's life. Without the encouragement from friends and colleagues as well as the support from my family, I would never be able to finish this book. They are all greatly appreciated.

After release of the first edition, I received positive feedback from many readers who appreciated this book for making the complex issues in rock mechanics simple. Many also provided constructive comments and suggestions for improvement and additional contents. This revised edition has taken all those suggestions into consideration and corrected known typing errors. A new chapter on application of numerical modelling in mining and geomechanics has also been added. I want to thank all of those who have provided constructive comments. My special thanks goes to Cui Lin for technical reading and Michelle Y. Zou for editing and proofreading.

Table of Contents

Preface

A real life story

A young rock mechanics graduate was on the way to work on his very first day, sharing a car with the Chief Mine Engineer and the Mine Superintendent of a mining company. As they drove along the highway at 6:00 am in the morning in northern Ontario, everybody was quiet. The Mine Superintendent started a conversation by asking the new fellow:

"Hi, Steve. You are a rock mechanics engineer. A mechanic can fix my car when it's broken. Can you fix the broken rock in our mine?"

Never expecting a question like that from his boss even before getting to his office, the young fellow hesitated for a few seconds, wondering what to say. Better to be honest, he thought and answered seriously: "No, a rock mechanics engineer cannot fix broken rocks. Instead, he crushes rocks and then tries to figure out why they were broken and how it happened." What he referred to was rock testing.

Perhaps, the boss was just making conversation with the engineer to be, hoping to help everyone relax in the half hour long car ride. Well, you can imagine how relaxed this young fellow would be.

The point is, an auto mechanic and a rock mechanics engineer are two completely different professionals. They are not doing the same thing. They use different approaches to solve different problems.

An auto mechanic deals with equipment, repairs and replaces broken parts. Once the work is done, the equipment may work as new.

A rock mechanics engineer, on the other hand, deals with rocks but cannot repair or replace already failed rock in the ground. He can however find a way to avoid or reduce the possibility of failure before the rock breaks in other areas or in similar conditions, using a completely different approach to solve problems.

That is what rock mechanics is about! Actually, there is more.

About this textbook

This book attempts to answer questions like the one above. It addresses various issues relevant to rock mechanics and its applications in the field. It is meant to be simple and easy to use so as to help readers who have little practical experience to understand the concepts and yet also provide practical solutions.

This book is intended for use by engineering students as a textbook. It may also be used by practical engineers and other professionals working in rock engineering as a quick reference on methodologies and design principles. The author attempts to balance fundamental knowledge and practical applications. This book first covers basics in rock properties, elemental stress analysis methods and rock failure mechanisms. It then discusses practical applications, such as engineering classification of rock masses, stress redistribution caused by underground excavations, ground-support interactions, design principles of rock structures, and ground support methods.

The content of this textbook covers the common knowledge and practice used in rock engineering. Examples are provided after introduction of a concept or a method. Practical applications are used to demonstrate their applications. Details of some aspects such as theoretical derivation, empirical formulation, etc. are omitted and only the results or the methodologies are presented in this book. Users who are interested in the origin of a theory, a design method, or formulation of empirical equations are encouraged to read relevant publications.

This revised edition has corrected the known typo and graphical errors in the previous edition, considered feedback from readers, and made changes in some difficult to understand topics to help readers. The revised edition has included more practical examples and also introduced new topics on numerical modelling and its application in rock engineering, which are considered very important and have become part of the routine work for practical engineers.

It is essential to point out that this book is by no means complete in covering all aspects of rock mechanics and its applications. Users are recommended to read other relevant articles, and updates of empirical methods, especially when nova design methods are developed. The author tries to present the complex issues in a simple way. However, users are suggested to have adequate knowledge in engineering mechanics and engineering geology.

Fair dealing with copyrighted materials

A textbook is supposed to present the common knowledge, design principles and methodologies accepted in practice. In engineering, the knowledge and methodologies are developed and refined over many years of practice by many practitioners. Many of the concepts, data and methodologies presented in this book are based on the works of other scholars and practitioners. To make it easier for readers to learn, the author tries to present the content in a simple way based on his own interpretation and use simple examples to demonstrate the applications. At the same time, the author endeavours to walk on a fine line between copyrighted materials and fair dealing. In cases like empirical design methods, the original design charts may be omitted and only extracted data and end results are presented in the book. Whenever possible, the original references are provided in the book. Users are suggested to consult with the original publications for detail. Instructors who intend to use this book in a class should arrange through their institutions to make those original materials available to their students.

Chapter 1

Introduction

1.1 Rock as We Know It

Rock is everywhere on Earth. It may outcrop on the surface or be covered by a layer of soil. What do we know about rocks?

In engineering geology, we learned about the structure of the earth's crust, rock formation, various types of rocks and geological structures in a rock mass.

In rock engineering, we look at rocks from a different perspective than other people. In comparison to geologists, we focus more on the mechanical aspects of rocks. In rock engineering design, we need to consider the factors that matter in ground stability.

There are many practical questions: Is rock a hard, unbreakable solid material? Is there any weakness in the rock? What is the load bearing capacity of a piece of rock? What influences the bearing capacity? How much stress is there in the ground? What affects the stability of the rock once an opening is excavated inside the rock mass?

To answer those questions, first, we need to become familiar with rocks from an engineering point of view. Rock appears as a solid mass (Fig. 1.1). It is in fact not perfect and has numerous imperfections, which can range from invisible micro-fractures, small fissures or bedding planes, to large joints or faults. These features can generally be called discontinuities or weaknesses in the rock masses. Some typical rock weaknesses, as seen in the field, are shown in Figs 1.2 to 1.5.

It is these weaknesses that need our attention. They often play dominant roles in ground instability and must be taken into consideration in rock engineering design.

To deal with these problems, we need knowledge in rock mechanics and ground stability control. However, the first challenge is to identify them and determine their spatial positions when they are not exposed to the surface. This requires knowledge in geological engineering and field mapping.

Fig. 1.1 Folding structure in rock, Sichuan, China 2010.

Fig. 1.2 Rock slope showing steep joints, Huidong, China 2008.

Fig. 1.3 Subsurface "sugar cube" rocks, Nova Scotia, Canada 2008.

Fig. 1.4 Rock slope with multiple joint sets, Sichuan, China 2009.

Fig. 1.5 Bedding structure in rock, MingJiang, China 2007.

1.2 History of Rock Engineering and Modern Civilization

Knowledge of rock mechanics is vital for geotechnical and mining engineering. However, it is only since the early 20th century that rock mechanics has come to be recognized as a discipline and as a topic of specialized lectures in an engineering program. That recognition is an inevitable consequence of engineering activities in rock, including complex underground installations, deep cuts for spillways and enormous open pit mines.

In the 18th and 19th centuries, large tunnels were driven for mine ventilation and drainage, water supply, canals, highways and railways. At present, rock work still plays a very important role in mining, petroleum and construction industries around the world.

Today, a multitude of important constructions on the surface, such as bridges, high-rise buildings and hydro dams, are established directly on top of rocks. Many others, such as tunnels, underground mines, hydro power stations, subway transit systems and underground storage spaces, are actually completed directly within the rock mass below the surface. Let us call these types of excavated space "**rock structures**". Each type of rock structure is intended for a specific use.

Such tremendous undertaking of rock structures is not risk free. Often ground failure occurs in various scales, such as foundation failure in geotechnical constructions, slope failure in open pit mines, ground failure and rockbursts in underground mines. There are also many examples of natural disasters such as landslides, earthquakes, sinkholes, subsidence and ground instability in other engineered projects. All of these problems must be dealt with using knowledge of rock mechanics.

Knowledge of rock mechanics, accumulated from experience in various rock engineering projects over the past two centuries, has gradually built up the special discipline in engineering that it is today. Methods unique for rock engineering have been developed for rock testing, field data collection, in-situ stress measurement, stability analysis, rock structure design and ground support.

1.3 Rock Mechanics as a Foundation of Rock Engineering

Either in dealing with rock structures in an engineering project, such as mining, slope and road cuts, tunnelling and foundation, or in dealing with natural disasters, such as landslide, subsidence, sinkholes and earthquake damage, it is essential for us to have knowledge of the rock on the site, the in-situ and the induced stresses, and how the rock would react to excavation. Rock mechanics is the foundation of rock engineering.

Rock mechanics deals with the mechanical properties of rocks and rock masses and uses special methodologies for design of rock-related engineering projects. Rock, somewhat similar to soil, is sufficiently distinct from other engineering materials that the process of "design" in rock is special. The close "cousin" in engineering is perhaps geotechnical engineering. In dealing with concrete structure, for example, an engineer first estimates the external load on the structure, prescribes the materials based on the required strength, and accordingly determines the structural geometry.

In dealing with rock structures, on the other hand, it is not the applied loads, but the existing stresses in-situ and the stresses induced by excavation that are important. These stresses often are unknown to us, both in magnitude and orientation. In addition, stresses underground are in three dimensions, which further complicates the problem. We will get back to that point later. Furthermore, the designed rock structures possess many possible failure modes. Determination of rock "strength" requires collection of geotechnical information in the field and then as much judgment as measurement. Finally, the geometry of rock structures to be designed is often limited, sometimes altered, by geological structures. the designing engineer may not have complete freedom.

For the above reasons, rock mechanics includes unique aspects: geological selection of sites rather than control of material properties, collection of geological and geotechnical data, measurement of in-situ stresses, and analysis through graphics and model studies, etc. The subject of rock mechanics is closely allied with geology and geological engineering.

What is rock mechanics?

There are various definitions. It is however more important to understand what is involved. In general, rock mechanics can be simply defined as an engineering branch, which:
- deals with rock materials,
- studies the mechanical properties of rocks,
- gives engineering consideration to rock masses in the field,
- applies engineering principles in design of surface and underground rock structures,
- furthermore, provides guidance to an engineer in design of mine openings, tunnels and road cuts.

It is important to note that rock engineering is unique in comparison with other engineering branches. Take structural engineering as an example, there are significant differences.

In structural engineering, the load and stress on the structure can be fairly well defined or estimated, the properties of the material (either concrete or steel) are known fairly well when they are made and the geometry of structures can be designed accordingly following standard practice.

In rock engineering, however, there are numerous unknowns and uncertainties in the field. The stress is not applied but exists in-situ, which is unknown in general. The stress will change if an opening is excavated in the rock mass. The strength of rock and rock masses varies widely from location to location. Rock formation varies too. Often, there are weak geological structures (e.g., joints, bedding planes, faults, shear zones, etc.) which weakens the rock mass in various degrees depending on their properties and orientations. All of these field factors are often unknown in advance and need to be "discovered" or measured, which is a difficult task as they are hidden in the rock masses underground.

These uncertainties pose a big challenge to an engineer and require unique approaches to solve engineering design problems. As a rock mechanics engineer, one should be proud of him-/herself for the special expertise in solving this type of ill-defined problems.

Rock mechanics versus soil mechanics

It is necessary at this point to make a brief comparison between rock mechanics and soil mechanics, the two seemingly close engineering branches.

Are they the same? No. Although rocks are the origin of soil, they are two different types of geo-materials and the engineering environment is different as well. Soil mechanics deals with soil on the surface or at shallow depth, subjected to a very small in-situ stress. Rock mechanics on the other hand deals with rocks from the surface to very deep in the ground, often subjected to high stresses in three dimensions. They both face uncertainties in the field as mentioned before. However, the variation encountered in rock mechanics is much wider. Usually what is used in one location may not be simply applied to another location even if the rocks appear the same.

From an engineering design point of view, there are more differences between soil and rocks as two different materials as shown below:

	Permeability and Seepage	Compressibility	Consolidation	Settlement
Soil:	varies	varies	loose ~ well	varies
Rock:	None or little	none	solid	none

Applications of rock mechanics

There are many areas where rock mechanics has applications. The following are a few examples of engineering projects where rock mechanics is applied.

In underground mining, often a mine may be surrounded by rock failure problems: roof fall, pillar failure, drift collapse or being squeezed, floor heaven, etc. These problems need to be solved. Even if a mine is safe without rock failure, one can still do more to optimize the design and reduce operation costs if they have a better idea of what they are dealing with.

In open pit mining, it is often the primary task for an engineer to optimize the design of the slope angle to balance the slope stability and excavation cost.

In road cuts and tunnelling, it is usually needed to design the engineering rock structure at a given geometry (size and shape) required for the intended operation, to estimate the anticipated ground failure and to provide ground reinforcement if necessary, taking costs into consideration.

1.4 Dealing with Uncertainties and Tolerance of Error

As indicated before, many factors, such as in-situ stresses, rock properties, orientations of weak planes, etc., required for design are often unknown in rock engineering. Field measurements will provide sample data and they can be used for design if they reasonably represent the conditions on the site. It is a fact that all measurements do not necessarily generate the same data and there are variations. Statistically, the more measurements, the more reliable the data are. The average of a large quantity of data on the same parameter should give reliable information, right? Wrong, or may-be, depending on what you are dealing with.

For example, the orientations of a number of joints are mapped in the field as shown in Fig. 1.6. It can be observed that there are two groups: one pointing to the east (90°) and another to the west (-90°) with some variations.

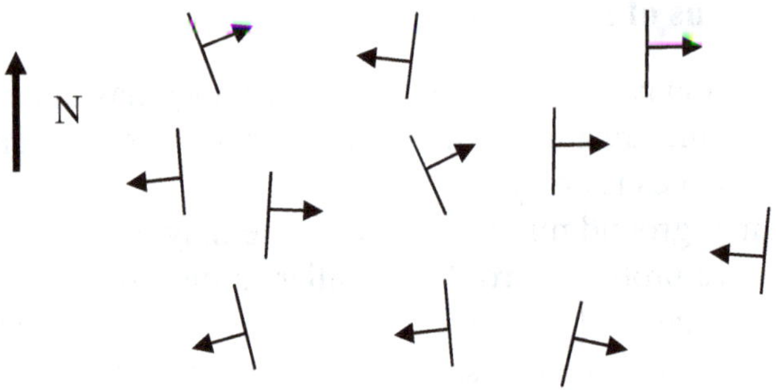

Fig. 1.6 Orientations of measured joints.

Suppose that the number of measurements is the same for each group. If the average dip direction of all joints is calculated, the result will be nearly zero, meaning it points North. This answer is obviously wrong and unacceptable! A special method is used to separate the two groups of data first.

Another example is measurement of in-situ stresses. In a simplified two-dimensional case, the in-situ stresses can be represented by the two principal stress components σ_1 and σ_2. If a circular opening is excavated in the rock, stresses around the opening are changed in both magnitude and orientation, as shown in Fig. 1.7. If a number of measurements are made in two symmetrical positions at equal distance from the center, an average of the magnitude of a stress component, such as σ_1, may be acceptable for the new stress condition around the opening. However, taking an average of their orientations is absolutely unacceptable. Stress state at each location after excavation is different and they cannot be simply added together. The in-situ stresses can only be calculated through relationships between the new stresses and the original stresses.

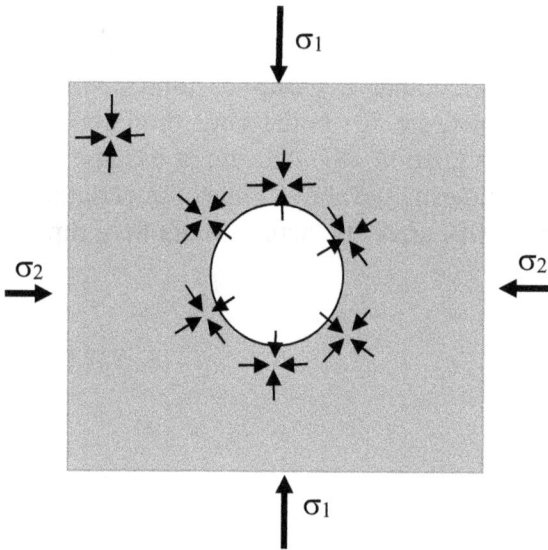

Fig. 1.7 Changes of stress magnitude and orientation after excavation.

Now, suppose that a number of measurements are made for the same group of the same factor, statistic averaging may be applied. The results will be an average plus a variance. For example, stress orientation of 45°±2° gives a range from 43° to 47°. That means any data in this range is acceptable. The "right" answer is not necessarily a single number, but in a range.

Engineering is about numbers and calculations. Mathematically, the addition of any two numbers will only give one answer:

e.g., 3.1 + 1.9 = 5.

Nothing more or less is accepted.

In rock engineering, addition of two measured numbers,

e.g., 3.1 + 1.9 = ?

The answer mathematically is still 5. However, when it comes to applying the result with a tolerable error (for example 4%), any value between 4.8 and 5.2 may be acceptable. That range of tolerance is to cover the uncertainties. Due to many uncertainties in the field, an error of 10% or more is very common and a range covering 10% - 15% may need to be considered in design and decision-making. If we can control the error within 10%, we may be doing an excellent work.

To apply the above concept, here is another example: Field mapping indicated that a group of joints, which are nearly parallel to a slope, dip 50° in the same direction as the slope. If the slope is designed at 48°, we cannot assume no effect from the joints, considering the above issue of uncertainty and the fact that within a 10% error the actual joints may dip between 45° and 55°.

Chapter 2

Behaviour of Rock

Note: *At low stress levels and room temperature, many materials are basically linearly elastic. However, they may deviate from elastic behaviour at high stress levels, elevated temperature or under a prolonged loading period.*

How would rock react to an external force? It depends on a number of factors including the internal properties of the rock itself and the loading conditions.

Any material deforms under an external load. For example, under compression it becomes shorter along the loading direction and expands laterally, as shown in Fig. 2.1. A plot of load P against deformation δ indicates how the material reacts to the load.

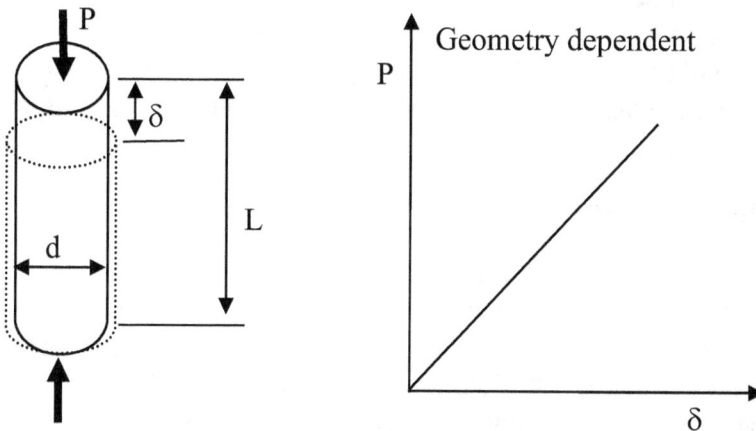

Fig. 2.1 Rock sample deformation under uniaxial loading.

For a simplified case, the P - δ relationship can be expressed as

$$P = k\delta \qquad (2.a)$$

where k is the slope, a constant called material stiffness.

It can be demonstrated that k increases with the size of the sample in Fig. 2.1. This would make it difficult to compare one material with another because of the size difference. To avoid this problem, the stress-strain σ - ε relationship is used instead.

Stress σ is defined as the load on a unit area. If the load P is applied evenly on the sample of diameter d, the induced stress is calculated by

$$\sigma = P/A \qquad (2.b)$$

where A is the cross-sectional area of the sample, $A = \pi d^2/4$ for this example.

Dimension of stress: [force/length^2].

Unit of stress: Pa ($=N/m^2$), kPa and MPa.

Strain ε, corresponding to the stress, is defined as the deformation per unit length

$$\varepsilon = \delta/L \qquad (2.c)$$

Dimension of strain: none; unit: none.

It should be noted that the σ - ε relationship is often not a straight line for most material in real life and the real relationship cannot always be described by the above simple form. This is particularly true for rocks because of their internal characteristics. Before we study the behaviour of rocks, we will start with idealized materials.

Idealized Materials

In the study of material behaviour, it is necessary to develop mathematical relationships that can relate stress, strain, deformation, rate of change of stress, rate of change of strain, or rate of change of deformation. This is usually based on reasonable assumptions. For idealized materials, the relationship between two or more of the above variables can be simplified. Various mathematical and mechanical models have been proposed to describe the mechanical behaviour of idealized materials. In the pre-failure stage, for example, the stress-strain relationship may be characterised as linear elastic, perfectly plastic, or a combination of the two. Some may also be time-dependent. In the following, the typical behaviour of idealized materials will be discussed before real materials, such as rock, is considered.

2.1 Linearly Elastic Materials

Under uniaxial compression, the stress-strain, σ - ε, relationship for a perfectly elastic solid material is demonstrated in Fig. 2.2 as a straight line. It can be written as

$$\sigma = E\varepsilon \qquad (2.1)$$

where E is the slope, a constant called Young's Modulus. This type of behaviour can be simulated by a spring model as shown in Fig. 2.2, which is useful in physical and numerical modelling.

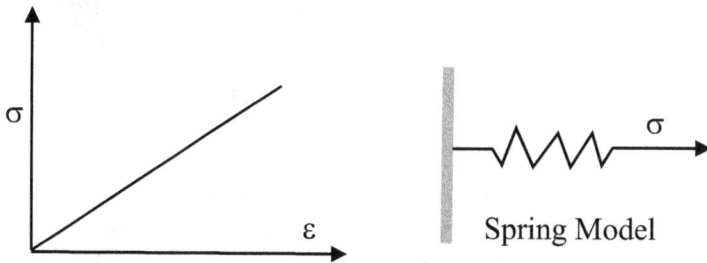

Fig. 2.2 Linear elastic model and its σ - ε relationship.

The following can be observed for a linear elastic material:
- the stress-strain σ - ε curve is a straight line through the origin,
- deformation increases linearly with the applied load,
- deformation is fully recoverable, i.e., the deformation will completely disappear after removal of the external load.

However, in reality no such material exists, although some igneous rocks do behave in a similar way.

2.2 Perfectly Plastic Materials

When a material shows no deformation before the stress reaches some level, and the ability of the material to support load is limited to that level, this type of behaviour is called perfectly plastic, as demonstrated in Fig. 2.3. The stress-strain σ - ε relationship is given as

$$\sigma = \sigma_0 \qquad\qquad (2.2)$$

where σ_0 is a constant, depending on the material property.

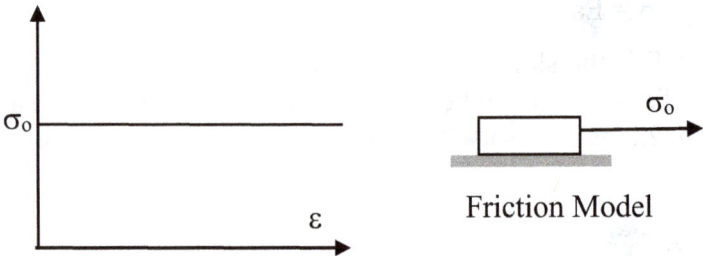

Friction Model

Fig. 2.3 Perfectly plastic model and its σ - ε relationship.

This means that the strain is not relevant to the stress once stress reaches σ_0. This type of behaviour can be simulated by a frictional surface-contact model as shown in Fig. 2.3.

For a perfectly plastic material, the following can be observed:

- the stress-strain σ - ε curve is a straight horizontal line,
- no deformation exists, i.e., $\varepsilon \equiv 0$, if $\sigma < \sigma_0$,
- deformation is permanent and non-recoverable, if $\sigma \geq \sigma_0$,
- no stress greater than σ_0 can be supported.

Again, in reality, no such material exists. There is more or less some deformation under external loading.

2.3 Elasto–Plastic Materials

The Venant material is perfectly elastic before stress reaches certain level and is perfectly plastic when stress reaches that level, as demonstrated in Fig. 2.4. This type of behaviour can be simulated by a spring-friction model.

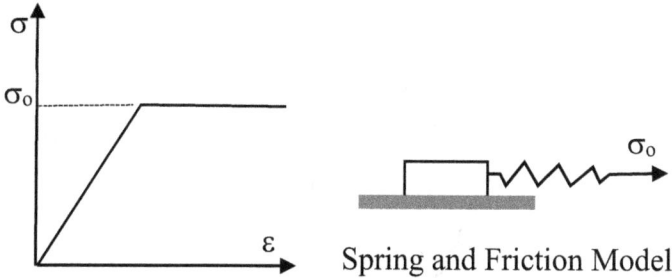

Spring and Friction Model

Fig. 2.4 Elasto-plastic model and its σ - ε relationship.

The stress-strain σ - ε relationship for elasto-plastic material is given as

$$\sigma = \begin{cases} E\varepsilon, \text{ if } \sigma < \sigma_0 \\ \sigma_0, \text{ if } \sigma \geq \sigma_0 \end{cases} \qquad (2.3)$$

For an elasto-plastic material, the following can be observed:
- the stress-strain σ - ε curve is a bisection straight line,
- it is linearly elastic, if $\sigma < \sigma_0$,
- it is purely plastic, if $\sigma = \sigma_0$.

The elasto-plastic model is a combination of the linear elastic and perfectly plastic models. It is an idealized situation of reality and more realistic than the previous two types of behaviour for some rocks.

2.4 Brittle and Ductile Failure

Rocks as we encountered in engineering projects, for example in mining and other types of underground excavations, will fail when they are overly loaded. The failure may be localized, from a small scale such as spalling, to a larger scale such as cracking and complete collapse. At this point it is also important for us to understand how a material is going to fail and what happens after the failure point. According to the behaviour of a material at and after failure, there are two types of failure: brittle failure and ductile failure, as demonstrated in Fig. 2.5.

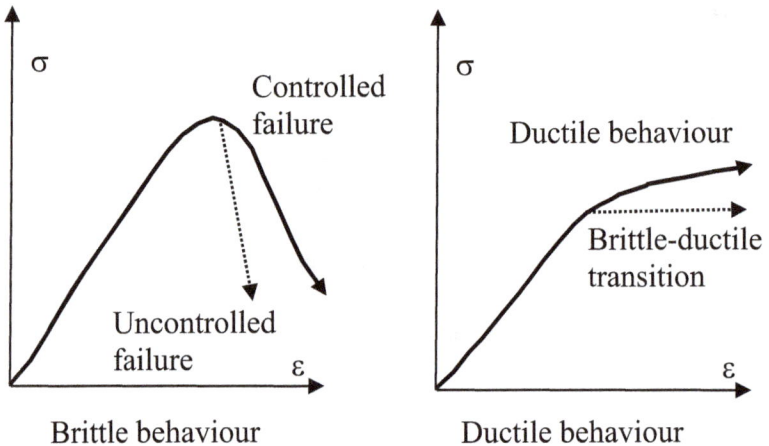

Fig. 2.5 Brittle – ductile behaviour under compression.

In the case of brittle failure, it can be observed that:
- the ability of the material to resist load after the failure point decreases with increasing deformation. Brittle failure is often associated with little or no permanent deformation (i.e., non-recoverable deformation) before failure, and,
- failure may occur suddenly, or catastrophically, depending upon the test/loading conditions. Rock bursts in deep hard rock mines provide good examples of the phenomenon of explosive brittle fracture. This feature is very important in hard rock mines because hard rocks tend to be brittle in nature.

In the case of ductile failure, it can be observed that:
- the material can sustain permanent deformation without losing its ability to resist load.
- ductility increases with increased confining pressure and temperature.

Most rocks will behave in a brittle rather than a ductile manner at the confining pressures and temperatures encountered in civil and mining engineering applications. Ductile behaviour can occur in weathered rocks, heavily jointed rock masses and some weak rocks such as evaporites under normal engineering conditions.

The mode of brittle or ductile failure is not absolutely fixed for a type of material and it may change when the environment changes. For example, as the confining pressure increases, some rock behavior will reach the *brittle-ductile transition* point at which there is a transition from typically brittle to fully ductile behaviour. Byerlee (1968) has defined the brittle-ductile transition pressure as the confining pressure at which the stress required to form failure plane in a rock specimen is equal to the stress required to cause sliding on that plane.

It should be noted that the brittle failure occurring in rocks under either laboratory or field conditions is often of a violent or uncontrolled nature. In this case, explosive failure occurs at the peak stress, and the post-peak section of the stress-strain curve cannot be recorded under normal conditions. In other situations, such as in mine pillars, the rock may be fractured and deform under a controlled manner beyond its peak load bearing capability and eventually reach equilibrium at some lower load. In such case, progressive fracture of the rock will be observed, and the post-peak section of the stress-strain curve can be recorded.

2.5 Typical Rock Behaviour

Now the question is: what kind of behaviour does a rock have under compressive load and how will it fail? There is no unique answer and none of the above will completely reflect rocks. Instead there is a mixture of them with rocks' own features. Without going through all the details of rock mechanics and the fracture mechanics (Griffith 1924) involving rock failure, typical rock behaviour in compression is presented in Fig. 2.6.

In general, typical rock behaviour includes the following stages:

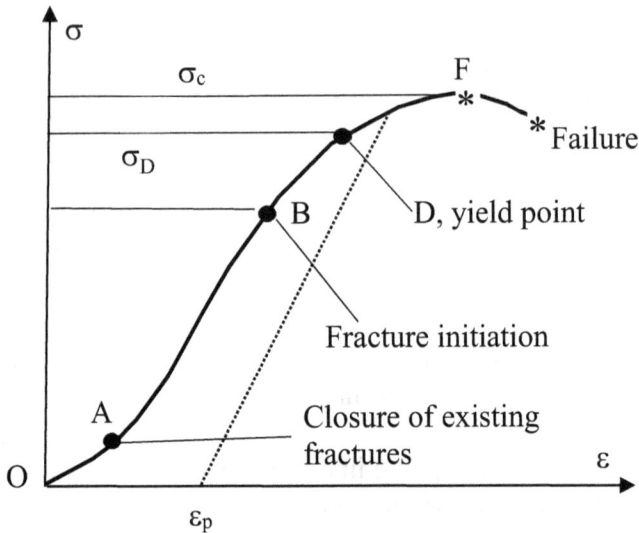

Fig. 2.6 Typical rock behaviour in compression.

- OA: closure of existing micro-fractures. Duration of this stage may be very short for hard and brittle rocks.
- AB: considered as linear elastic and deformation is fully recoverable.
- BD: nonlinear elastic and stable fracture propagation. During this stage, the propagation stops if external load is removed.
- DF: ductile, un-stable fracture propagation. Beyond the yielding point D, permanent deformation develops and the deformation ε_p is non recoverable even if the load is completely removed. At this point, the rock is said to have developed a "permanent set". During this stage, fracture propagation may not stop even if the external load is removed. For some rocks, especially brittle ones, DF may be very short or may not exist at all.

At point F, the allowable stress reaches the maximum limit, i.e., the material strength. This is the uniaxial compressive strength σ_c under uniaxial compression. Failure is expected to occur right after point F.

Beyond Point F is the post-failure stage. How this stage is going to develop will depend on the type of material and the conditions of loading, and whether failure is brittle or ductile as discussed in the previous section. If the post-failure is controlled, such as on a servo-controlled stiff testing machine or for a highly confined rock in the field, the full stress – strain, σ - ε curve can be recorded.

In practice, rock behaviour in the field is often represented by a simplified behaviour model as shown in Fig. 2.7. Ignoring the first stage of fracture closure, which is reasonable for most hard and brittle rocks, the rock pre-failure stage is represented by a straight line for linear elastic behaviour and the post-failure stage by a plastic model with a reduced strength σ_r.

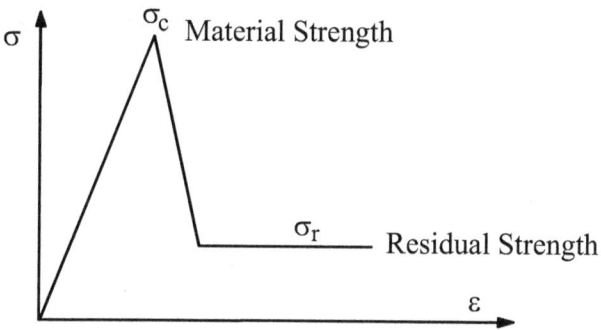

Fig. 2.7 Simplified model of rock behaviour.

The above simplified rock behaviour model is suitable for practical applications, such as numerical modelling and supports design.

2.6 Effects of Confining Pressure

The behaviour of a material is well known to be affected by the loading conditions, e.g., the confinement, temperature, loading rate, etc. It has been known qualitatively for a century that if the lateral displacement of a specimen in a compression test is resisted by applying pressure to its sides, it becomes stronger and there is a tendency toward greater ductility. There have been numerous studies on this subject in the past (for example, Griggs

1936). In general, if a confining pressure is applied to the sides of a specimen as in triaxial tests, or if there is confinement to the rock mass as in the field, the stress-strain σ - ε curve is expected to change as shown in Fig. 2.8.

In Fig. 2.8, the lowest curve is from uniaxial loading. The confining pressure σ_3 resists the deformation in the lateral direction and therefore changes the rock behaviour significantly. As the confining pressure increases,

- the σ - ε curve changes from brittle to ductile gradually,
- the failure point also moves higher which means higher strength,
- the permanent set increases, but remains small,
- the ductility increases, a very important feature for rocks confined at great depth.

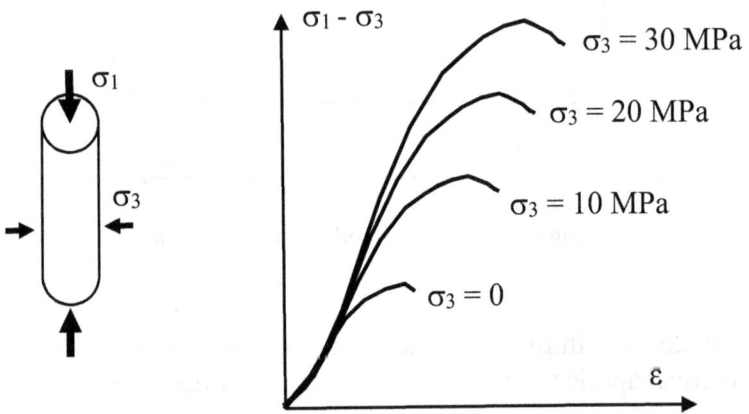

Fig. 2.8 Illustration of the effect of confining pressure on rock behaviour.

2.7 Effects of Temperature

High temperature is expected to cause a molecular effect in the rock. Detailed discussion of the effect is out of the scope of this textbook. It should however be understood.

The effect of temperature on rock behaviour is also well studied (for example, Griggs et al. 1960) and it has become a known fact that increasing the temperature lowers the brittle-ductile transition pressure. Figure 2.9 illustrates the temperature effects for a granite sample at a confining pressure of 500 MPa. At room temperature the behaviour is brittle, but at higher temperatures substantial amounts of permanent deformation may be introduced without loss of load. At 800°C the material is almost fully ductile.

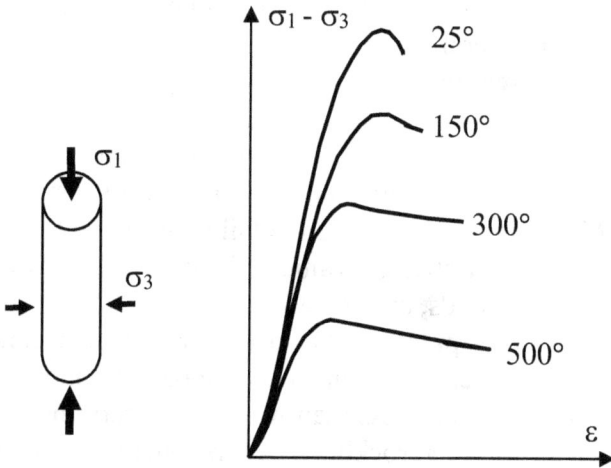

Fig. 2.9 Illustration of the effect of temperature (°C) on rock behaviour.

The brittle-ductile transition may be studied as a function of both pressure and temperature. In general, as the temperature increases, the strength decreases and the ductility increases.

The brittle-ductile transition is of great interest to researchers involving rock in deep ground or a high temperature environment, but it is of less interest to rock engineers since the temperature in a work environment is limited to a relatively narrow range with much less variation. The temperature effect may not be so important as that of confining pressure.

2.8 Effects of Loading Rate and Time Dependency

Effect of Loading Rate

In the above discussions, the results are normally from static loading, i.e., the load is increased from zero gradually to a certain level within a period of time, which can be a few minutes or longer. The actual rate of loading may also have some effect on the results. This effect is of less importance to static loading in comparison with dynamic loading, in which case the load is applied instantly to a level. Loading rate directly affects the strain rate. In general, as the loading rate increases, the recorded compressive strength increases and the transition pressure from brittle to ductile may change as well.

The change of strength can be surprisingly big! For example, in a laboratory test, a sandstone showed a compressive strength $\sigma_c = 56$ MPa when it was loaded to fail in 30 seconds. The same material indicated a strength value of 84 MPa when it was loaded to fail in 0.03 seconds, an increase of 50%.

In engineering practice, the loading rate that occurs in the field is much lower than that in the laboratory and the actual strength of rock can be less than what was determined in a test. The exception is when rockburst occurs where the "loading" to the surrounding rock mass is instant.

Time Dependency

In a laboratory test, if different rock samples are loaded to a certain stress level and the stress is kept constant for some time, you may find that some of the rock samples would fail at a later time. That is the result of creep! If a rock displays creep characteristics, the deformation will continue and even increase with time under a constant stress when the stress has reached a particular level. Studies on creep of rock go back to as early as the 1930s (for example, Griggs 1939). Figure 2.10 illustrates idealized creep curves of salt.

It can be seen in Fig. 2.10 (right chart) that when stress is very low, the creep effect may not appear. As the stress reaches a certain level, which depends on the type of rock, the creep effect may become significant. Once a material displays the creep effect, it develops in three phases (Fig. 2.10, left):

Phase I: transient phase, strain will increase quickly before becoming steady

Phase II: steady-state, strain increases linearly with time

Phase III: tertiary phase, strain increases sharply again prior to failure.

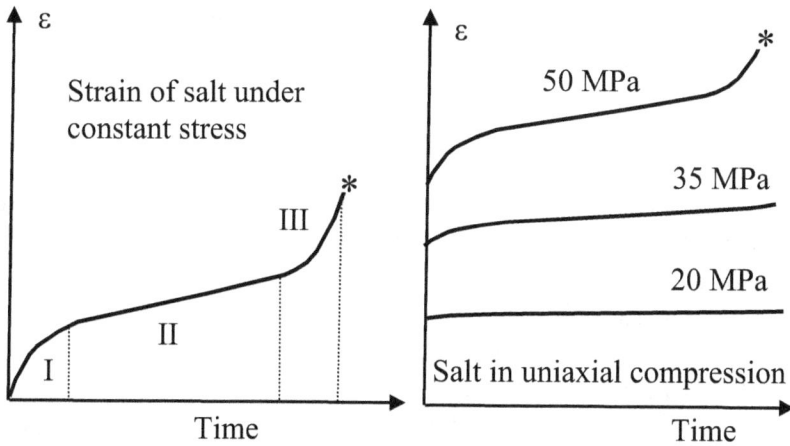

Fig. 2.10 Illustration of time dependence and the effect of stress level on the strain of salt in constant compression.

The duration of the three phases and the magnitude of change in each phase vary with the material properties. The duration of Phase III is usually very short and once it is initiated, failure can not be arrested. Most rocks creep. However, soft rocks creep much more than hard rocks.

Viscous models are normally used to describe creep behaviour. In general, the creep curve for a number of materials can be expressed by

$$\varepsilon = a + b\,t + c(f) \tag{2.4}$$

where a is the elastic strain at time t = 0, b is a constant, bt is the creep developed during the steady-state and $c(f)$ is the transient creep. Tertiary creep is not included because of its short duration before the impending failure.

As a rock creeps under constant load, its strength may also decrease. A stable condition may become unstable after some time if the stress is close to the strength in magnitude. Both strain increase and strength decrease will have negative impact on the stability of rock structures.

Creep in the Field

In relation to the design of rock structures, the duration of the transient creep period is often too short to consider. However, immediately following an underground blast there is evidence of the effects of transient creep as manifested by small rock falls and development of fractures, which generate both audible and micro-seismic noise. These effects usually decrease rapidly with time so that 15 to 60 minutes after blasting the disturbed area returns to normal. This afterworking immediately following blasting presumably relieves high stress concentration on the newly created rock surfaces.

Steady-state creep can be measured over long periods in the roof and pillars of underground mines. Figures 2.11 and 2.12 illustrate the typical time dependency of roof convergence and pillar expansion, respectively.

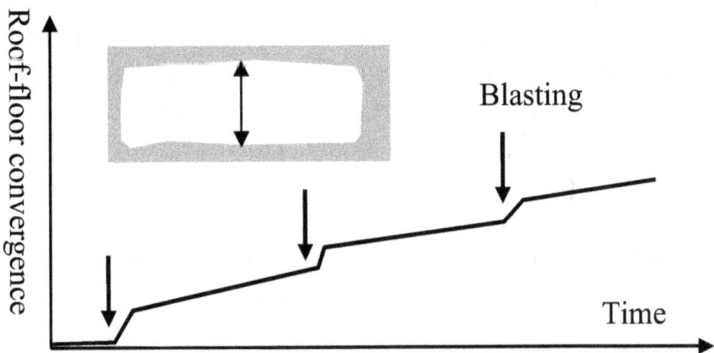

Fig. 2.11 Illustration of time dependence of roof convergence in soft rocks.

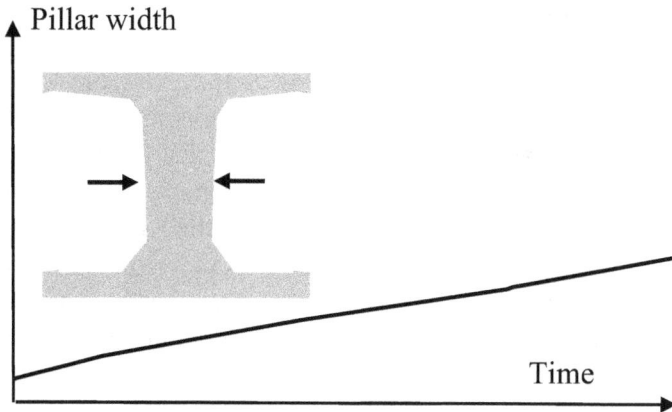

Fig. 2.12 Illustration of time dependence of pillar expansion in soft rocks.

From monitoring data, it is possible to identify the blasting induced movement (as marked by arrows in Fig. 2.11) and the creep effects. In the field, such measurement located in strategically important points in a mine can help engineers understand what is happening in the rock mass.

The creep rate may also depend on the rock temperature. From an engineering standpoint, for hard rocks this effect is usually negligible within the range of temperatures encountered in mining. However, creep in salt, potash and other evaporite minerals is temperature dependent and in some cases this effect may be a primary factor in design considerations. More creep data on salt can be found in Jeremi (1994).

2.9 Anisotropy of Rocks

In the above discussions, the loading direction (the line along which a force is applied) is not considered. When the loading direction is changed, the behaviour of a rock may be very different and the strength can change significantly if there is a weak joint in the rock sample. Variation of compressive strength with respect to the direction of the principal stress is termed "strength anisotropy". Strong anisotropy is characteristic of rocks composed of parallel arrangements of flat minerals such as mica, chlorite, and clay or long minerals such as hornblende.

Thus, the metamorphic rocks, especially schist and slate, are often markedly directional in their behaviour. Anisotropy also occurs in regularly interlayered mixtures of different components as in banded gneisses sandstone-shale alternations, or chert-shale alternations. In such rocks, strength varies continuously with direction and demonstrates pronounced minima when the planes of symmetry of the rock structure are oblique to the major principal stress. Rock masses cut by joints also display strength anisotropy, except where the joint planes lie nearly perpendicularly to the major principal stress direction. Strength anisotropy is illustrated in Fig. 2.13, where the relative direction of a joint with the principal stress σ_1 is represented by an angle Ψ.

As shown, a joint has the most significant impact on the strength of the sample when the joint plane has an angle Ψ near 30° from the maximum stress direction. It basically has no effect at 60° or more (i.e., a joint nearly perpendicular to the maximum stress). More discussion will be given in a later chapter.

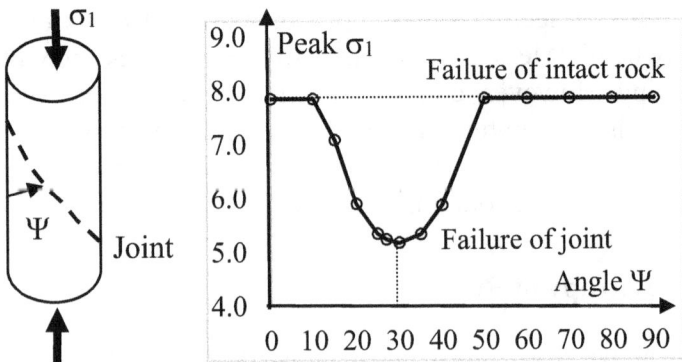

Fig. 2.13 Effect of joint direction on rock strength.

Chapter 3

Mechanical Properties of Rock

We understand that underground rocks are generally subjected to a load equal to or greater than the weight of the overlying rock. The reaction to load is different for different types of rocks. This reaction can be described by the load-displacement, or stress-strain relationships as discussed in previous chapters. This type of qualitative description is not easy to put into engineering practice. Some sort of engineering indices or measurable parameters are required to measure quantitatively the rock properties so that we can apply them in actual engineering design. Some of the basic and commonly used mechanical property indices are discussed below.

3.1 Stiffness

Stiffness, k, is defined as the slope of the P - δ curve as shown in Fig. 2.1. From Eqn. 2.a, k can be determined as

$$k = P/\delta \qquad (3.1)$$

Dimension of k: [force/length].
Unit of k: [N/m].

Stiffness k is useful in indicating the ability of a rock pillar or a specimen to resist external load. It is dependent upon the size (or width) of the pillar: k $\propto$ size. When pillar size increases, k increases. Therefore, it does not reflect the ability of the rock and is not comparable for different types of rock.

3.2 Young's Modulus

Young's Modulus, E, is defined as the slope of the σ - ε curve as shown in Fig. 3.1 and is given by the stress/strain ratio

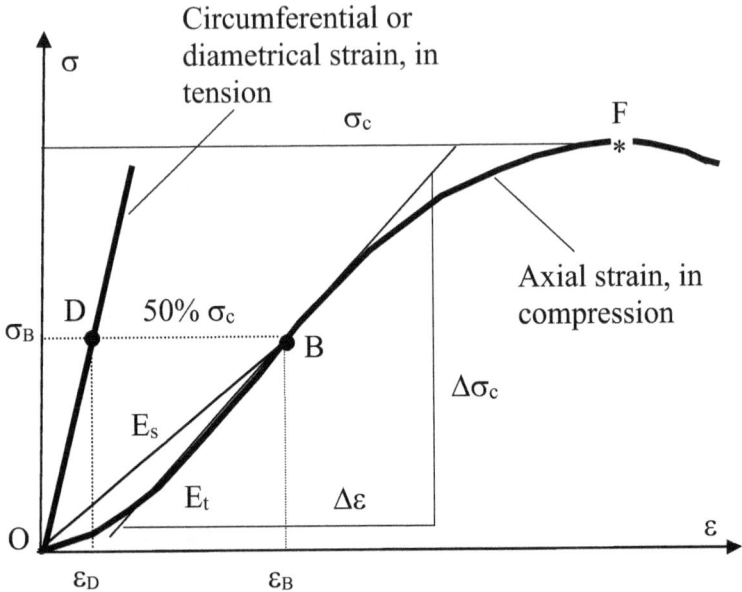

Fig. 3.1 Typical stress – strain curves in uniaxial compression.

$$E = \sigma / \varepsilon \tag{3.2}$$

Dimension of E: [force/length^2].
Unit of E: normally, [GPa] for rocks.

As the σ - ε curve is usually not linear, the slope varies depending on the stress level on the curve. In Canada, for the purpose of design, modulus and Poisson's ratio, which will be discussed later, are calculated at a stress level (Point B) corresponding to 50% of the uniaxial compressive strength σ_c.

Depending on the way the slope is drawn, two types of modulus can therefore be defined: tangent modulus E_t and secant modulus E_s. Tangent Modulus is the slope of a straight line tangent to the σ - ε curve at point B and is determined by

$$E_t = \frac{\Delta \sigma}{\Delta \varepsilon} \tag{3.3}$$

where $\Delta\sigma$ and $\Delta\varepsilon$ are the stress- and strain-increments, respectively as shown in Fig. 3.1.

Secant Modulus, E_s, is the slope of a straight line connecting point B and the origin O and is determined by

$$E_s = \frac{\sigma_B}{\varepsilon_B} = \frac{0.5\,\sigma_c}{\varepsilon_B} \qquad (3.4)$$

Normally, $E_s \leq E_t$ $\qquad\qquad\qquad\qquad\qquad\qquad$ (3.5)

For hard rocks with no or few micro-fractures (nearly linear elastic), the OA portion is very small on the σ - ε curve in Fig. 2.6 and $E_s \approx E_t$. In general, the higher the modulus, the stronger the rock is. The value of the modulus varies widely for different types of rock, ranging from < 1.0 GPa to 90 GPa or much higher. For coal, this value is typically between 0.7 and 7 GPa and for rocks between 7 and 91 GPa.

It should be noted that under dynamic loading, the value of E is higher than that determined under static loading.

3.3 Poisson's Ratio

Poisson's ratio υ is defined as the ratio of the lateral strain over the axial strain and calculated at the stress level of 50% σ_c.

$$\upsilon = \frac{\varepsilon_D}{\varepsilon_B} \qquad (3.6)$$

where ε_B is the axial strain along the loading direction and ε_D is the lateral strain perpendicular to the loading direction. ε_D could be the diametrical strain or the circumferential strain for a cylindrical specimen as indicated in Fig. 3.1.

Poisson's ratio is dimensionless and has no unit. The larger the value of υ, the more expandable the material is. The value of υ is generally in the range of 0.06 to 0.45, with a maximum of 0.5 for elastic materials.

However, it should be noted that if a rock sample contains joints with rough surfaces, slip along joints may cause dilation perpendicular to the joint planes. As a result, the whole sample may expand more than expected, resulting in a much higher value of the measured expansion ratio (Singh and Singh 2008, Song et al. 2015). This is in fact the effect of the discontinuity, not the rock itself. This is referred to as the volumetric expansion ratio of jointed rock, υ_j and is not to be confused with Poisson's ratio υ.

3.4 Uniaxial Compressive Strength

Uniaxial compressive strength (UCS), σ_c, is the maximum stress on the stress-strain curve before the material fails under one dimensional loading as demonstrated in Fig. 2.1. σ_c is calculated by Eqn. 2.b at the failure point and has the same dimension and unit as stress. The two terms: stress and strength, must not be confused, although they both have the same unit and are calculated by the same equation.

Some typical values of rock strength from laboratory tests are as follows:

Coal, 7 to 56 MPa
Shale,   14 to 140 MPa,
Sandstone, 49 to 240 MPa,
Granite,   up to 280 MPa,
Quartzite,   up to 450 MPa or higher.

All the three parameters E, υ, and σ_c can be determined from a single uniaxial compression test if strain gauges are attached to the rock specimen along the loading direction and perpendicular to that direction.

Example 3.1

Given the σ - ε curve recorded in a laboratory test, as shown on the next page. Determine: E_s, E_t, υ and σ_c.

Solutions:

1). From the top of the curve, $\sigma_c = 125$ MPa.

2). At the point of 50% of σ_c, $\sigma = 62.5$ MPa, $\varepsilon_a = 0.0031$.
   By Eqn. 3.4,
        $E_s = 62.5$ MPa / $0.0031 = 20.2$ GPa.

3). Draw a tangent line to the σ - ε curve at 50% σ_c

$\Delta\sigma = 62.5 - 14 = 48.5$ MPa
$\Delta\varepsilon = 0.0031 - 0.00175 = 0.00135$
By Eqn. 3.3,
$E_t = 48.5$ MPa $/ 0.00135 = 35.9$ GPa.

4). At the point of 50% of σ_c, $\varepsilon_a = 0.0031$ and $\varepsilon_D = 0.0007$,
By Eqn. 3.6,
$\upsilon = \varepsilon_D/\varepsilon_a = 0.0007/0.0031 = 0.226$.

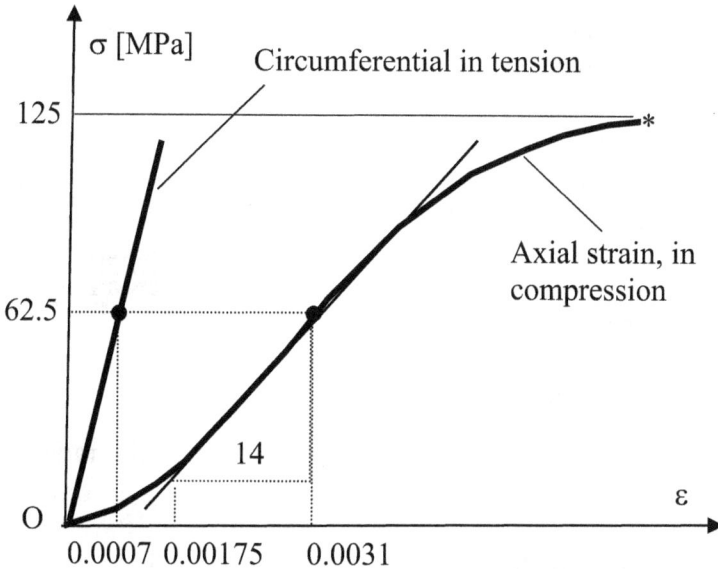

(σ - ε curve for Example 3.1.)

3.5 Tensile Strength

Tensile strength, σ_t, is the maximum allowable tensile stress a material can stand prior to failure. When a rock specimen is under tension, it should fail on a plane where it has the least resistance to tensile stress. For a homogeneous isotropic material, the plane of failure should be perpendicular to the tensile stress.

In general, the tensile strength of rock is very low, $\sigma_t \leq 10$ to 20 MPa for intact rock and will be much lower if a weak plane exists. In most cases, $\sigma_t \approx 0$ is assumed for jointed rock masses for design purpose in rock engineering.

σ_t can be determined by two different methods: the direct pull test and the indirect compression test.

A. Direct Pull Test:

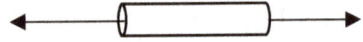

In this test, a pull force is applied to the rock specimen and a tensile stress is induced in the specimen. If there is a weak plane in the specimen, failure is most likely going to take place along that weak plane. σ_t is calculated by Eqn. 2.b except that P is a tensile force at failure.

Although the direct pull test seems logical and straight forward, this method is very difficult to implement on rocks. The challenge is in applying a pull force to the specimen's ends considering the brittle nature of rocks. Therefore, an alternative method is usually used.

B. Brazilian Indirect Test:

In this test, a disk specimen with a length/diameter ratio of 0.5, i.e., $t \approx 0.5d$ in Fig. 3.2, is used. In the test, a compressive distributed force is applied to the specimen along the diametrical central plane as shown in Fig. 3.2. The contact between the loading device and the specimen should be in a small area rather than a single line to avoid local crushing. This can be achieved by wrapping the specimen with a layer of masking tape.

Based on elasticity theory, under the compressive force, pure tension is developed along the diametrical central plane and failure is caused by a tensile stress on that plane.

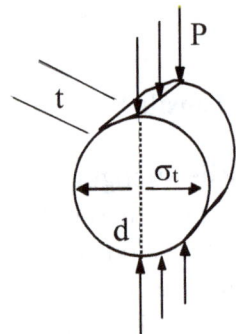

Fig. 3.2 Brazilian test.

The tensile strength is calculated by

$$\sigma_t = \frac{2\,P}{\pi\,d\,t} \tag{3.7}$$

This method is simple and is widely used. The result of σ_t from the Brazilian test is usually equal to or higher than that from the direct pull test. A calculation example is given in Appendix II.

3.6 Shear Strength on Rock Surfaces

Shear strength on a rock surface is one of the most important factors in slope stability analysis and in underground fault slip analysis. A potential failure surface may consist of a single discontinuity plane or a complex path following several discontinuities involving fractures of the intact rock material. Determination of reliable shear strength values is a critical part of slope design.

Shear strength, τ_s, on a rock surface is defined as the maximum shear stress required to cause slip failure (i.e., slip or slide movement) along the surface. This is referred to as the peak shear strength and can be determined by a direct shear test as illustrated in Fig. 3.3.

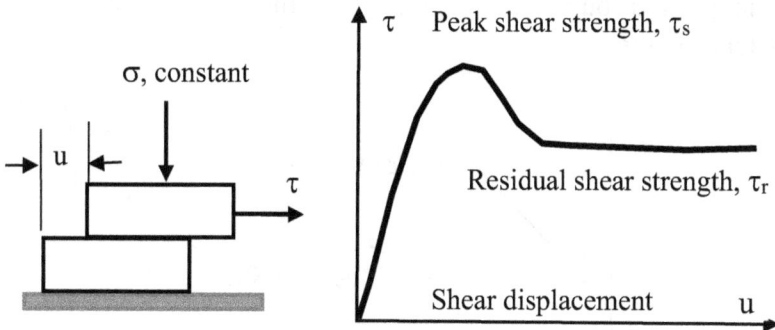

Fig. 3.3 Illustration of a direct shear test of rock surface.

Under a constant normal stress, a shear stress is applied to the upper block and increased gradually. At the same time, both shear displacement, u and shear stress, τ are recorded. A plot of the data is also shown in Fig. 3.3.

As can be seen, initially shear displacement increases with the shear stress. Once the shear stress has passed a certain value, it will no longer increase. Any attempt to increase the shear stress will only accelerate the displacement. The shear stress required to maintain the shear movement drops to a lower level. The stress corresponding to the highest point on the τ - u curve is the peak shear strength, τ_s and thereafter is the residual shear strength, τ_r. The joint shear stiffness can be determined on the τ - u curve, $\Delta\tau/\Delta u$, i.e., the slope of the curve before slip starts.

It is noted that the shear strength of a joint surface, or discontinuity is much lower than that of a saw-cut surface in the intact rock. Rock mass in the field usually contains one or more discontinuities. In rock engineering, it is often the discontinuity which controls the ground stability.

The values of the peak and the residual shear strengths depend on the normal stress acting on the rock surface. If a series of tests are carried out on the same type of rock surface under different normal stress levels, a set of results of τ_s and τ_r can be obtained. With these data and the corresponding normal stresses, two strength curves can be plotted on the τ - σ plane, as shown in Fig. 3.4. It needs to be noted that the residual shear strength line goes through the origin, while the peak shear strength intersects the τ axis.

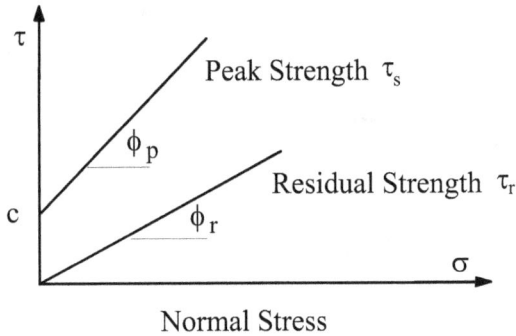

Fig. 3.4 Shear strength versus normal stress on a rock surface.

The τ value at the point where the peak shear strength line intersects the τ axis is called the cohesion, c. The slope angles of the two lines are called the peak friction angle, ϕ_p and residual friction angle, ϕ_r, respectively. They have the following relationship:

$$\phi_r \leq \phi_p \qquad (3.8)$$

The two shear strength lines can be expressed as:

$$\tau_s = c + \sigma \tan \phi_p \qquad (3.9)$$

$$\tau_r = \sigma \tan \phi_r \qquad (3.10)$$

where c is a constant and has the same unit as strength.

It should be pointed out that shear failure may take place through a plane other than a joint surface in the rock mass. The shear strength on such a plane is normally much higher than that on a joint plane and it can be determined by the triaxial compressive test, which will be discussed later in the triaxial test section.

Influence of water pressure

Pressure of water, p, in a discontinuity has negative impact on rock strength. The primary influence of water is a reduction of shear strength due to a reduction of the *effective* normal stress as a result of water pressure. This effect can be incorporated into the shear strength equation by using effective normal stress. Dropping the subscripts and re-writing Eqns. (3.9 and 3.10) in a unified form, we have

$$\tau = c + \sigma' \tan \phi \qquad (3.11)$$

where σ' is the effective normal stress,

$$\sigma' = \sigma - p \qquad (3.11a)$$

Eqn. 3.11 can be applied to both peak and residual shear strength if ϕ is replaced by ϕ_p or ϕ_r (at c = 0) correspondingly.

Presence of water may also affect the filling or cementing materials in a rock discontinuity and in turn influence the shear strength. In most hard rocks and in many sandy soil and gravels, these properties are not significantly altered by water but many clays, shales, mudstones and similar materials will exhibit significant changes as a result of changes in moisture content. It is important, therefore, that shear tests are carried out on samples which are as close as possible to the in-situ moisture content in the rock mass.

Influence of surface roughness

It is common knowledge that the shear strength on a rock surface will vary with the surface roughness. The rougher the surface, the higher the shear strength. The surface roughness of a rock surface has a complex profile under a microscope. There have been numerous studies on how to measure and account for the roughness in determination of shear strength (e.g., Barton 1976 and 1989, Patton 1966).

Figure 3.5 illustrates simplified models of surface roughness, where the unevenness, or small bumps and ripples on the rock surfaces are represented by a roughness angle, i. Three types of rock surface profiles are shown in the figure: stepped, undulating and planar.

Previous research indicated that at a very low normal stress, these small bumps come into play and the value of i is added onto the basic friction angle ϕ,

$$\tau = c + \sigma' \tan (\phi + i) \tag{3.12}$$

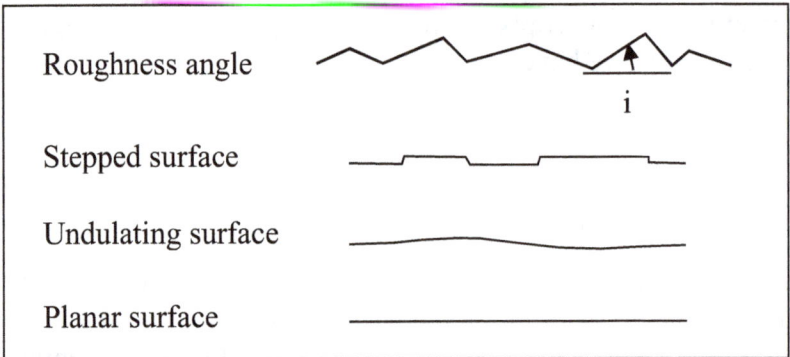

Fig. 3.5 Illustration of rock surface profile and surface roughness.

As the normal stress increases, these bumps will be sheared off (i.e., $i \Rightarrow 0$) and the large undulation will take effect.

Another approach to account for the surface roughness is to use Barton's empirical equation (1973),

$$\tau = \sigma \tan [\phi + JRC \log_{10} (\sigma_c/\sigma)] \qquad (3.13)$$

where use of ϕ_r is suggested and JRC is the joint roughness coefficient, depending on the roughness of the rock surface. The cohesion is ignored because it is relatively small in comparison.

Estimate of JRC

In the field, the value of the joint roughness coefficient (JRC) can be determined by inspection of the joint surface conditions and comparison with the suggested values (Barton 1987), which are summarized in Table 3.1 for convenience. The same roughness may have a different effect depending on the scale of the test. The suggested JRC values at a 200 mm scale in Table 3.1 is for laboratory tests and at a 1.0 m scale for field estimates.

JRC can also be estimated from a simple tilt test in which a pair of matching discontinuity' surfaces are tilted until one slides on the other. The JRC value is

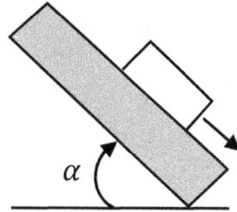

Table 3.1 Summary of JRC values (After Barton 1987).

Surface Condition	JRC (200 mm scale)	JRC (1 m scale)	Equivalent J_r
Stepped, rough	20	11	4
Stepped, smooth	14	9	3
Stepped, slikensided	11	8	2
Undulating, rough	14	9	3
Undulating, smooth	11	8	2
Undulating, slikensided	7	6	1.5
Planar, rough	2.5	2.3	1.5
Planar, smooth	1.5	0.9	1.0
Planar, slikensided	0.5	0.4	0.5

estimated from the tilt angle α by the following equation:

$$JRC = \frac{\alpha - \varphi}{\log_{10}(\sigma_c/\sigma)} \qquad (3.14)$$

where both α and ϕ are given in degrees, and σ is the normal stress on the discontinuity surface.

For small samples, the normal stress σ may be as low as 0.001 MPa. As an example, assuming this normal stress value in a typical case in which the tilt angle $\alpha = 65°$, the basic friction angle $\phi = 30°$ and the joint wall compressive strength $\sigma_c = 100$ MPa, Eqn. 3.14 gives JRC = 7.

The applicable range of Barton's Eqn. 3.13 is limited to use at low normal stress: $0.01 < \sigma/\sigma_c < 0.3$ and the value of $[\phi + JRC \log_{10}(\sigma_c/\sigma)]$ in Eqn. 3.13 is limited to $\leq 70°$. In most slope stability problems, the normal stress falls within this range. Therefore, this equation is sufficient to use for slope stability analysis.

Example 3.2

Tests of rock samples give a compressive strength $\sigma_c = 150$ MPa and the frictional angle on a joint surface $\phi = 30°$. The joint surface is relatively smooth and slightly undulating.

Determine the shear strength by Barton's empirical equation at an estimated normal stress of 40 MPa.

Solutions:

From Table 3.1, use JRC = 11 for an undulating smooth surface and lab test sample (use 8 if for field estimation). By Eqn. 3.13

$\tau = \sigma \tan[\phi + JRC \log_{10}(\sigma_c/\sigma)]$

$= 40 \text{ MPa} \tan[30° + 11 \log_{10}(150 \text{ MPa} / 40 \text{ MPa})]$

$= 40 \text{ MPa} \tan(30° + 6.31°)$

$= 40 \text{ MPa} \times 0.735$

$= 29.4 \text{ MPa}$ (make sure all units are compatible).

3.7 Triaxial Compressive Strength and Internal Shear Strength

Underground rock structures are normally under confinement. As shown in Chapter 2, confining pressure in the triaxial test will affect the behaviour of rock and its strength. Similar to the uniaxial compressive strength, the triaxial compressive strength is simply determined at the highest point of a σ - ε curve, i.e., the axial stress σ_1 at the failure point at a corresponding confining stress level σ_3. If a series of tests are conducted, each at a different confining pressure level, a number of σ - ε curves can be plotted as shown in Fig. 2.8 and the corresponding strength values can be calculated.

As described above, the triaxial strength is not a single value but a series of data: σ_1 at a given σ_3. The higher the σ_3, the higher the σ_1 will be. If the data of (σ_1, σ_3) is plotted on a σ_1 - σ_3 plane, a strength envelope can be drawn as shown in Fig. 3.6. When $\sigma_3 = 0$, $\sigma_1 = \sigma_c$. The strength envelope separates the stable zone ($\sigma_1 < \sigma_{1i}$) from the unstable/failure zone ($\sigma_1 > \sigma_{1i}$). σ_{1i} is considered as the instantaneous compressive strength at a given confining stress σ_3.

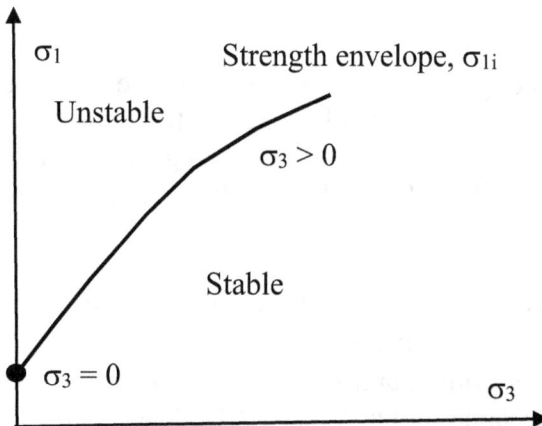

Fig. 3.6 Triaxial strength envelope.

Internal shear strength and internal frictional angle

For each triaxial test, a Mohr circle (detail given in next chapter) can be drawn on the τ - σ plane. The circle is centered on the σ axis with σ_1 and σ_3 as two points across its diameter as shown in Fig. 3.7.

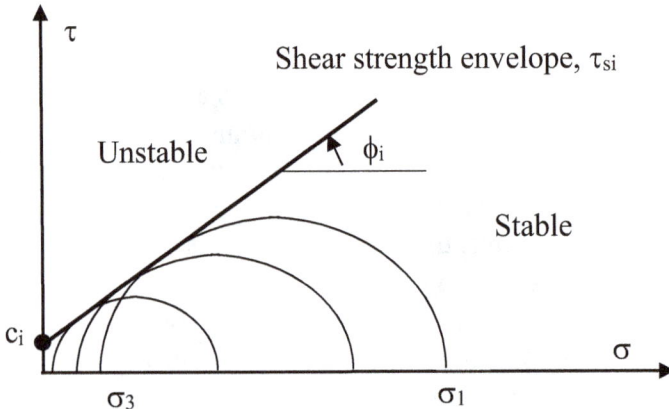

Fig. 3.7 Shear strength envelope from triaxial tests.
(τ_{si} is as the instantaneous shear strength at a given normal stress σ.)

The radius of the circle: $R = (\sigma_1 - \sigma_3)/2$,
The centre of the circle is at: $\tau = 0$ and $\sigma = (\sigma_1 + \sigma_3)/2$.

With the set of data shown in Fig. 3.6, a series of circles can be drawn along the σ axis. A best fit-curve (not necessarily a straight line) tangent to all circles will be the shear strength envelope. Within a certain range (typically a linear elastic range), a simplified straight line can be used as an approximation

$$\tau_{si} = c_i + \sigma \tan \phi_i \qquad (3.15)$$

where ϕ_i is the internal friction angle and c_i is the internal cohesion of the rock mass.

Normally both c_i and ϕ_i vary with the stress level σ. Eqn. 3.15 has the same format as Eqn. 3.11. Once again, the strength envelope separates the stable zone ($\tau < \tau_{si}$) from the unstable/failure zone ($\tau > \tau_{si}$).

3.8 Point Load Strength Index

This is an old, yet simple and inexpensive index to measure the rock strength. With this test method, a point load is applied on a specimen between two contact points until the rock fails, as shown in Fig. 3.8. The test can be done on cylindrical core samples or lump rock samples. Load can be applied parallel or perpendicular to the core axis. Figure 3.8 also indicates the loading direction and the requirements of sample dimensions.

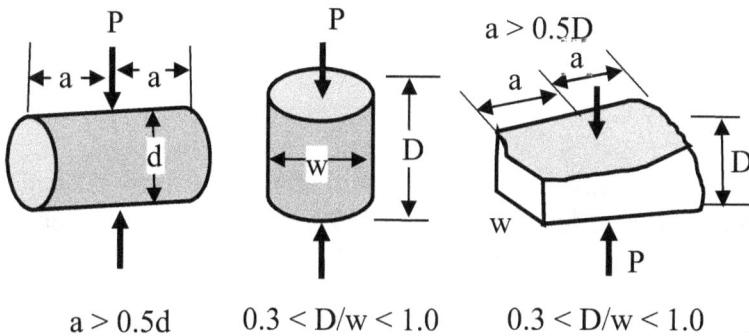

Fig. 3.8 Point load directions and measurements.

The Point load strength index is determined by

$$I_s = P / D_e^2 \tag{3.16}$$

where P is the load at failure, D_e is the equivalent diameter of the cross sectional area through platen contact points.

For diametrical test (left in Fig. 3.8),

$$D_e = d \tag{3.17a}$$

For other tests, De is defined as

$$D_e^2 = 4 \, w \, D / \pi \tag{3.17b}$$

where D and w are defined in Fig. 3.8.

The test can be done on any specimen size. However, the strength index is corrected for the size effect. The value of size

$d = 50$ mm is used, $I_{s(50)}$, as the standard. Correction can be done by a chart (commonly used before) or by the following formula:

$$I_{s(50)} = I_s (D_e / 50)^{0.45} \tag{3.18}$$

where De is in mm.

It should be noticed that I_s is not directly comparable with the uniaxial compressive strength σ_c because of the different loading conditions and failure mechanisms. However, since I_s is easy to determine even in the field, is inexpensive and serves the purpose, it is still used in practice today. There is an approximate empirical relationship between these two strength indices:

$$\sigma_c \approx 24 \ I_{s(50)} \tag{3.19}$$

Example 3.3

A rock core specimen with a diameter of 48 mm is loaded diametrically to failure at $P = 1200$ kg on a point load tester. Determine the $I_{s(50)}$ value of the rock.

Solution:

For diametrical loading, $D_e = d$ (Eqn. 3.17a).

$I_s = P/D_e^2 = P/d^2$

$\quad = 1200$ kg $\times$ 9.8 N/kg $/ (0.048$ m$)^2$

$\quad = 5,104,167$ N/m^2 – 5.1 MPa.

Correction for standard size at 50mm, by Eqn. 3.18

$I_{s(50)} = I_s (D_e / 50)^{0.45} = 5.1 \ (48 / 50)^{0.45} = 5.0$ MPa.

It is recommended to test at least 10 specimens and the average is calculated. An example is given in Appendix III.

Chapter 4

Elemental Stress Analysis

The stresses in the field are in a three-dimensional state. There are three principal stresses perpendicular to each other. Analysis of 3D stresses is a very complex issue. Fortunately, in many cases, the issues encountered in mining and geotechnical engineering can be reasonably dealt with as a 2D stress condition. Therefore, discussions below will focus on 2D stress analysis.

4.1 Stresses in Two Dimensions

Let us consider an infinitesimally small square rock element loaded by stresses in both x and y directions, as shown in Fig. 4.1.
  - coordinate system: xy
  - convention: all stresses shown in Fig. 4.1 are **positive**.
  - three independent stress components: normal stresses: σ_x and σ_y, and shear stresses:

$$\tau_{xy} = \tau_{yx} \tag{4.1}$$

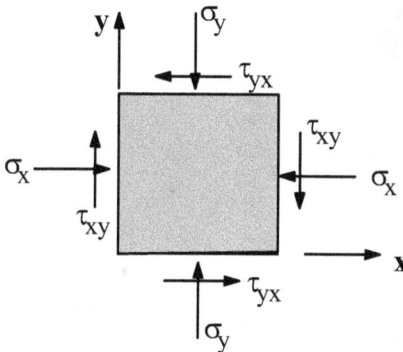

Fig. 4.1 2D Stress definition on an infinitesimal element.

Stresses on a plane or in a direction

Assume that a plane, which is perpendicular to the x-y plane, cuts through the element, leaving a trace AB as shown in Fig. 4.2. The normal of plane AB is a vector **n**, within the x-y plane and inclines at an angle θ from the x axis. θ is positive when measured counter-clockwise from the x axis. In this case, the angle between the tracing line of the plane AB and the y axis is also θ.

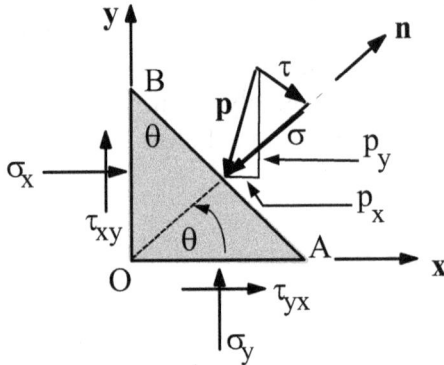

Fig. 4.2 Stress resolution in plane x-y.

If the forces acting on the element were in equilibrium and the element was at rest, all forces exerted by the stresses over the element after separation by plane AB must also be in equilibrium. In order to balance the stresses on the two sides OA and OB, there is a stress vector **p** acting on plane AB, **p** being restricted within the x-y plane. The stress vector **p** can be resolved in the x and y directions as two components, p_x and p_y. Thus,

$\Sigma F_x = 0$:

$$p_x \, AB = \sigma_x \, OB + \tau_{xy} \, OA$$

Because OB = AB cos θ and OA = AB sin θ

$$\therefore \quad p_x = \sigma_x \cos\theta + \tau_{xy} \sin\theta \tag{4.2}$$

Similarly, with $\Sigma F_y = 0$, we have

$$p_y = \sigma_y \sin\theta + \tau_{xy} \cos\theta \tag{4.3}$$

At the same time, the stress vector **p** can also be resolved in the normal and in-plane directions on plane AB, giving another two stress components, σ - the normal stress and τ - the shear stress acting on AB. Both σ and τ are restricted within the x-y plane and $\sigma \perp \tau$.

From the geometrical relationships in the stress diagram of Fig. 4.2, stress is balanced in the normal direction,

$$\sigma = p_x \cos \theta + p_y \sin \theta$$

Substituting Eqns. 4.2 and 4.3 in the above equation and simplifying, we have

$$\sigma = \sigma_x \cos^2\theta + \sigma_y \sin^2\theta + 2 \tau_{xy} \sin \theta \cos \theta \qquad (4.4)$$

Substituting $\cos^2\theta = (1 + \cos 2\theta)/2$ and $\sin^2\theta = (1 - \cos 2\theta)/2$ in Eqn. 4.4 and re-arranging the terms, we have

$$\sigma = \frac{\sigma_x + \sigma_y}{2} + \frac{\sigma_x - \sigma_y}{2} \cos 2\theta + \tau_{xy} \sin 2\theta \qquad (4.5)$$

Similarly, when stress is balanced in the in-plane (shear) direction,

$$\tau = - p_x \sin \theta + p_y \cos \theta$$

Substituting Eqns. 4.2 and 4.3 in the above equation and simplifying, we have

$$\tau = -(\sigma_x - \sigma_y) \sin \theta \cos \theta + \tau_{xy} (\cos^2\theta - \sin^2\theta) \qquad (4.6)$$

$$\tau = -\frac{\sigma_x - \sigma_y}{2} \sin 2\theta + \tau_{xy} \cos 2\theta \qquad (4.7)$$

Equations 4.5 and 4.7 allow us to calculate the normal and shear stress components $\{\sigma, \tau\}$ in the direction **n** (i.e., on the plane AB), in a stress field of $\{\sigma_x, \sigma_y, \tau_{xy}\}$.

4.2 Principal Stresses

Definition: If the shear stress is zero on a plane, the normal of which is **n**, then

- the normal stress on the plane is called the principal stress,
- the direction **n** is called the principal direction,
- the plane is called the principal plane,
- an axis parallel to the principal direction is called the principal axis.

Determination of principal directions and principal stresses

In Eqn. 4.7, let $\tau = 0$, we have

$$\text{Tan } 2\theta = \frac{\tau_{xy}}{(\sigma_x - \sigma_y)/2} \tag{4.8}$$

Solving Eqn. 4.8 we will have two answers:

$$\theta_a \in [0°, 90°] \text{ and } \theta_b \in [-90°, 0°).$$

Replacing $\sin 2\theta$ and $\cos 2\theta$ in Eqn. 4.5 with the following trigonometrical relationships:

$$\sin 2\theta = \pm \frac{\tan 2\theta}{\sqrt{1 + \tan^2 2\theta}}, \quad \cos 2\theta = \pm \frac{1}{\sqrt{1 + \tan^2 2\theta}} \tag{4.9}$$

and substituting $\tan 2\theta$ by Eqn. 4.8, we can solve for σ, arriving at the following two solutions:

$$\sigma_{1,2} = \frac{\sigma_x + \sigma_y}{2} \pm \sqrt{(\frac{\sigma_x - \sigma_y}{2})^2 + \tau_{xy}^2} \tag{4.10}$$

where if the operator $\pm$ is replaced by +, the result is σ_1 and if replaced by -, the result is σ_2.

In a two-dimensional stress field, there are two principal stresses and directions. σ_1 is called the major principal stress and σ_2 the minor principal stress, $\sigma_1 \geq \sigma_2$.

Equations 4.8 and 4.10 do not necessarily have corresponding answers. We need to go one step further to find the principal directions for σ_1 and σ_2. The easiest and simplest way is to substitute the value of θ_a or θ_b calculated by Eqn. 4.8 into Eqn. 4.5 and to calculate σ. If the obtained value matches σ_1, the corresponding θ is for σ_1, and vice versa.

In a special condition where σ_1 and σ_2 are parallel to x and y directions, respectively, as shown in Fig. 4.3, x becomes the major principal direction and y the minor principal direction. In this case, the stresses in θ direction have a simplified form and they can be derived from Eqns. 4.5 and 4.7 directly by replacing σ_x by σ_1 and σ_y by σ_2, and setting $\tau_{xy} = 0$.

$$\sigma = \frac{\sigma_1 + \sigma_2}{2} + \frac{\sigma_1 - \sigma_2}{2} \cos 2\theta \qquad (4.11)$$

$$\tau = -\frac{\sigma_1 - \sigma_2}{2} \sin 2\theta \qquad (4.12)$$

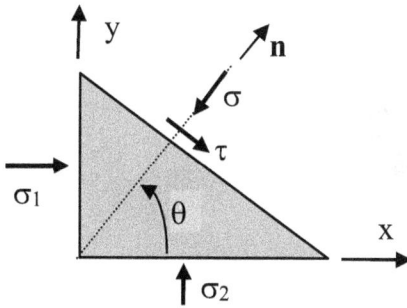

Fig. 4.3 Stresses related to principal directions.

Equations 4.11 and 4.12 indicate that both the normal and shear stresses are a sine function of angle θ. As θ changes, the shear stress varies between its maximum value (>0) and its minimum value (<0), and the normal stress varies between its major principal value (σ_1) and its minor principal value (σ_2), as shown in Fig. 4.4.

When θ =0, ± 90° or 180°, $|\sin2\theta| = 0$, the shear stress τ =0, and the normal stress reaches the value of the corresponding principal stress.

When θ = ± 45° or ± 135°, $|\sin2\theta| = 1.0$, the shear stress reaches its maximum magnitude, $|\tau|$ = the maximum. These are the maximum shear directions that define the maximum shear planes (**Note**: the direction **n** is perpendicular to a plane).

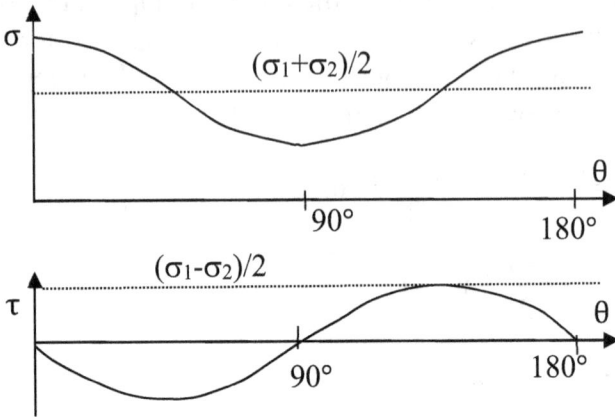

Fig. 4.4 Normal and shear stresses vary as a sine function of θ, θ being measured from the σ_1 direction.

On the maximum shear planes, e.g., when $\theta = \pm 45°$ from the σ_1 direction, the stresses are

$$\sigma = \frac{\sigma_1 + \sigma_2}{2}, \quad \tau = \mp \frac{\sigma_1 - \sigma_2}{2} \tag{4.13}$$

Example 4.1

Given $\sigma_x = 24$ MPa, $\sigma_y = 20$ MPa and $\tau_{xy} = 5$MPa.
What are
   a) the normal and shear stresses in the direction of $\theta = 30°$ from the x axis counter clockwise?
   b) the principal stresses?
   c) the principal directions (show them in a diagram)?

Solutions:

First draw a stress element diagram based on the given conditions to show all stresses specified. Then mark the direction of $\theta = 30°$ on the diagram, as shown by **n**.

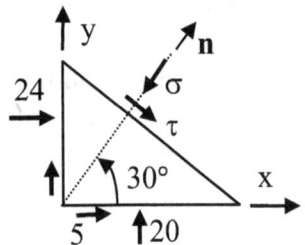

a). The given angle θ is positive as defined in Fig. 4.2.

By Eqn. 4.5,

$$\sigma = \frac{24+20}{2} + \frac{24-20}{2}\cos(2 \times 30°) + 5\sin(2 \times 30°)$$

$$= 22 + 1 + 4.33$$

$$= 27.33 \text{ (MPa)}$$

By Eqn. 4.7,

$$\tau = -\frac{24-20}{2}\sin(2 \times 30°) + 5\cos(2 \times 30°)$$

$$= -1.73 + 2.5$$

$$= 0.77 \text{ (MPa)}$$

b). By Eqn. 4.10,

$$\sigma_{1,2} = \frac{24+20}{2} \pm \sqrt{(\frac{24-20}{2})^2 + 5^2}$$

$$= 22.0 \pm \sqrt{29}$$

$\therefore \sigma_1 = 27.4$ (MPa) and $\sigma_2 = 16.6$ (MPa).

c). By Eqn. 4.8,

$$\tan 2\theta = \frac{5}{(24-20)/2} = 2.5$$

$\therefore \theta = \frac{1}{2}\tan^{-1}(2.5) = 34.1°$.

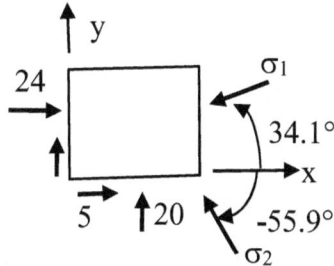

Substitute $\theta = 34.1°$ into Eqn. 4.5,

$$\sigma = \frac{24+20}{2} + \frac{24-20}{2}\cos(2 \times 34.1°) + 5\sin(2 \times 34.1°)$$

$$= 22 + 0.743 + 4.64$$

$$= 27.4 \text{ MPa}.$$

$\therefore \sigma_1$ direction is $\theta_1 = 34.1°$,

and σ_2 direction is $\theta_2 = 34.1° - 90° = -55.9°$.

See the above diagram for both directions.

Example 4.2

Given $\sigma_1 = 27.4$ MPa, $\sigma_2 = 16.6$ MPa, with σ_1 // (parallel) to the x direction. What are the normal and shear stresses on
   a) a plane which is $\perp$ (perpendicular) to the x-y plane and forms an angle of 55.9° with x axis,
   b) a plane which is $\perp$ to the x-y plane and the normal of which is 55.9° with x axis,
   c) the plane which is $\perp$ to the x-y plane and has the maximum shear stress?

Solutions:

a). Since the specified plane forms an angle of 55.9° from the x axis, the angle θ of the plane's normal is
   $\theta = 55.9° - 90° = - 34.1°$ as shown in the margin.
Since σ_1 is // to the x axis, the simplified Eqns. 4.11 and 4.12 can be used. By Eqn. 4.11,

$$\sigma = \frac{27.4 + 16.6}{2} + \frac{27.4 - 16.6}{2} \cos(-2 \times 34.1°)$$

$$= 22 + 5.4 \cos(-68.2°)$$

$$= 24 \text{ MPa}.$$

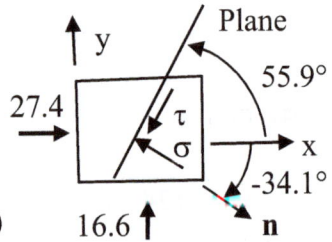
Plane
27.4
y
τ
σ
x
55.9°
-34.1°
16.6
n

By Eqn. 4.12,

$$\tau = -\frac{27.4 - 16.6}{2} \sin(-2 \times 34.1°)$$

$$= 5 \text{ MPa}.$$

b). Because the angle of the plane's normal is specified as $\theta = 55.9°$, by Eqn. 4.11 and results of a) above, we have
   $\sigma = 22 + 5.4 \cos(2 \times 55.9°)$
     $= 20$ MPa,

   $\tau = - 5.4 \sin(2 \times 55.9°)$
     $= - 5$ MPa.

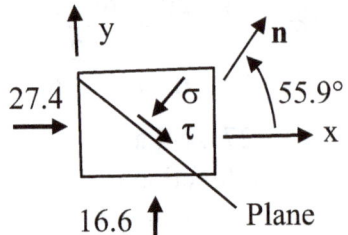
27.4
y
σ
τ
n
55.9°
x
16.6
Plane

c). The maximum shear stress occurs on the planes, the normal
of which is $\theta = \pm 45°$ from the σ_1 direction.

By Eqn. 4.13,

$\sigma = (27.4 + 16.6)/2 = 22$ MPa,

$\tau = \mp (27.4 - 16.6)/2$
$= \mp 5.4$ MPa.

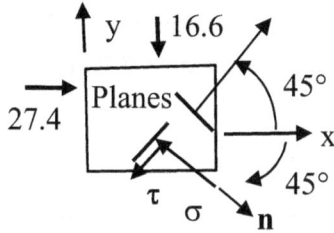

Example 4.3

Geological structures (faults, weak beddings, shear zones, etc.)
are weaker than the rocks that make up the rock mass in the field.
They play an important role in ground instability of engineering
constructions. Failure associated with these structures is usually
through shear underground and may involve separation in rock
slopes. A geological structure can be approximately considered
as a plane. The stresses in underground are eventually resolved
into two components acting upon the plane: a normal stress and
a shear stress. Depending on the position of a joint plane (dip
and dip direction), the two stresses acting on the plane may vary
significantly even in the same stress field.

Let us consider a horizontal drift underground in an xyz
coordinate system as shown below. A joint is traced on the drift
walls and the joint plane is found to
be parallel to the drift and dips 40°
towards the x direction.
Deformation on any cross section
of the drift, except near its ends, are
considered identical in an isotropic
and uniform rock mass. This is a
plane strain condition.

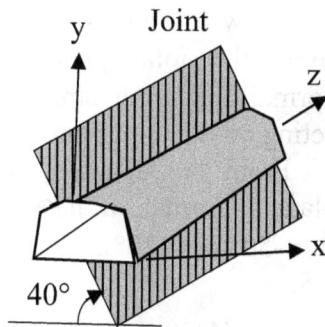

The stresses underground are in
three dimensions. Determination of
the stresses in the field is a very complex topic by itself and more
detail will be discussed in a later chapter. At this point, we
assume that the stresses acting on the drift cross section (the x-y

plane) are $\sigma_v = 10$ MPa and $\sigma_h = 14$ MPa without shear stress. Let us try to answer these questions:

  a) What are the normal and shear stresses on the joint plane?
  b) What are the normal and shear stresses on the joint plane if the joint dips towards the opposite direction?

Solutions:

a). First draw a diagram to show the in-situ stresses, the joint plane on the cross section. Then draw the normal and shear stresses (σ, τ) acting on the joint plane.

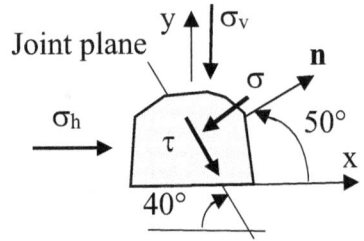

From the diagram, the joint plane's normal, **n** has an angle $\theta = 50°$.

From the given condition, $\sigma_x = 14$ MPa, $\sigma_y = 10$ MPa, $\tau_{xy} = 0$. Then by Eqn. 4.5,

$$\sigma = \frac{14+10}{2} + \frac{14-10}{2}\cos(2 \times 50°) + 0$$

$$= 12 - 0.35 + 0 = 11.65 \text{ (MPa)}.$$

By Eqn. 4.7,

$$\tau = -\frac{14-10}{2}\sin(2 \times 50°) + 0$$

$$= -1.97 \text{ (MPa)}.$$

b). Draw a diagram as above to show the joint plane and the normal and shear stresses (σ, τ) acting on the joint plane.

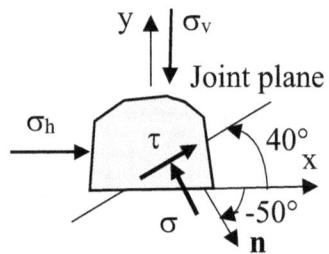

From the diagram, the joint plane's normal, **n** has an angle $\theta = -50°$ (or $130°$). Then by Eqn. 4.5,

$$\sigma = \frac{14+10}{2} + \frac{14-10}{2}\cos(-2 \times 50°) + 0$$

$$= 12 - 0.35 = 11.65 \text{ (MPa)}, \text{ the same value as in a).}$$

By Eqn. 4.7,

$$\tau = -\frac{14-10}{2}\sin(-2\times 50°)+0$$

= 1.97 (MPa), different value and direction from a).
(note: same results if $\theta = 130°$ is used.)

4.3 Mohr's Circle

Mohr's circle is a graphical method of stress analysis. It was a convenient engineering tool especially when computers were not yet available. It continues to be a useful tool today as it gives visual information about the stresses.

Mohr's circle is drawn on a σ - τ plane as shown in Fig. 4.5. If the principal stresses are known, σ_1 and σ_2 are the two diametrical points on the σ axis. The circle
- is centered on the σ axis at coordinates $((\sigma_1 + \sigma_2)/2, 0)$,
- has a radius $R = (\sigma_1 - \sigma_2)/2$.

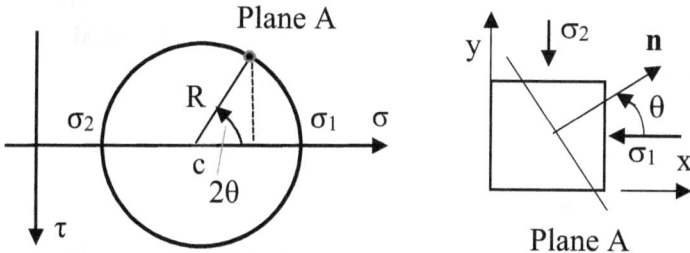

Fig. 4.5 Mohr's circle representation of stress state and stresses on plane A
(**Note**: τ axis is set to point downwards as positive)

As discussed before, a plane, which is $\perp$ to the x-y plane, is represented by the plane's normal direction, **n**. The following points must be considered when using Mohr's circle:
- The stresses (σ, τ) on a plane (or in a direction) is represented by the coordinates at a point on Mohr circle.
- The θ value is doubled (i.e., 2θ) on Mohr circle and is measured in the same direction on the Mohr circle as on the stress element diagram.

- The starting point for measuring θ (and 2θ) must be the same on both the Mohr circle and the stress element (e.g., σ_1 direction).

As demonstrated in Fig. 4.5, if a plane A's normal, **n** has angle θ measured counter-clockwise from σ_1 on the stress element. The stresses on plane A are represented by the coordinates at point A at 2θ, also counter-clockwise from σ_1 on the Mohr circle.

From the geometry of the diagram, it is easy to determine the coordinates (σ, τ) of point A - the stresses on plane A.

$$\sigma = \frac{\sigma_1 + \sigma_2}{2} + Ac \cdot \cos 2\theta = \frac{\sigma_1 + \sigma_2}{2} + \frac{\sigma_1 - \sigma_2}{2} \cdot \cos 2\theta \quad (4.14)$$

$$\tau = -Ac \cdot \sin 2\theta = -\frac{\sigma_1 - \sigma_2}{2} \sin 2\theta \quad (4.15)$$

These two equations are identical to the analytical solutions given earlier by Eqns. 4.11 and 4.12.

A Mohr diagram is an alternative method to analytical solutions for stress analysis. It can be used to solve any stress problems discussed in the previous section. The detail will be better explained by the following examples.

Example 4.4

Solve the problem given in Example 4.2 by Mohr's diagram:

Given $\sigma_1 = 27.4$ MPa, $\sigma_2 = 16.6$ MPa, with σ_1 in the direction of the x axis. What are the normal and shear stresses on

a) a plane which is ⊥ to the x-y plane and forms an angle of 55.9° with the x axis,

b) a plane which is ⊥ to the x-y plane and the normal of which is 55.9° with x axis,

c) the plane which is ⊥ to the x-y plane and has maximum shear stress?

Solutions:

Construction of Mohr's circle on graph paper:
- mark σ_1 at P and σ_2 at Q on the σ axis of the σ - τ plane,

- find the midpoint between P - Q,
- draw a circle through the two points centered at the midpoint.

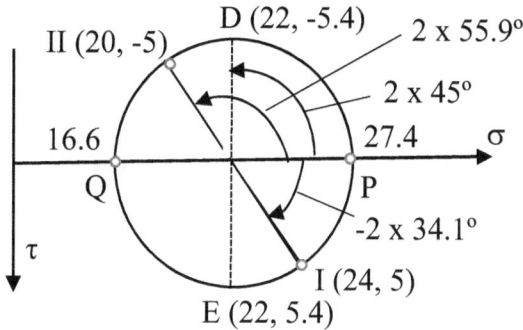

a). As explained in Example 4.2, the normal of the specified
plane has an angle $\theta = -34.1°$ from σ_1 (clockwise).
  - on the Mohr circle, measure $2\theta = -2 \times 34.1°$ from P (or σ_1)
  clockwise and mark it as Point I,
  - measure the coordinates of Point I:
  $\sigma = 24$ MPa, $\tau = 5$ MPa.

b). In this case, the normal of the specified plane has an angle
  $\theta = 59.9°$ from σ_1 (counter-clockwise).
  - on the Mohr circle, measure $2\theta = 2 \times 59.9°$ counter-
  clockwise from P and mark it as Point II,
  - measure the coordinates of Point II:
  $\sigma = 20$ MPa, $\tau = -5$ MPa.

c). The normal of the maximum shear plane is
    $\theta = \pm 45°$ from σ_1.
  - on the Mohr circle, measure $2\theta = \pm 2 \times 45°$ from P and mark
  them as Points D and E, respectively
  - measure their coordinates:

  at D ($\theta = 45°$), $\sigma = 22$ MPa, $\tau = -5.4$ MPa,

  at E ($\theta = -45°$), $\sigma = 22$ MPa, $\tau = 5.4$ MPa.

Example 4.5

Solve the problem given in Example 4.1 by Mohr's diagram.
Given $\sigma_x = 24$ MPa, $\sigma_y = 20$ MPa and $\tau_{xy} = 5$ MPa. What are:
 a) the normal and shear stresses in the direction of $\theta = 30°$
    from the x axis counter clockwise?
 b) the principal stresses?
 c) the principal directions (show them on a stress diagram)?

Solutions:

Construction of Mohr's circle (see below):
 - mark the x axis as I at (24, 5) and y axis as II at (20, -5),
 - connect Points I and II to find the centre c on the σ axis,
 - draw a circle through the two points I and II centered at c.

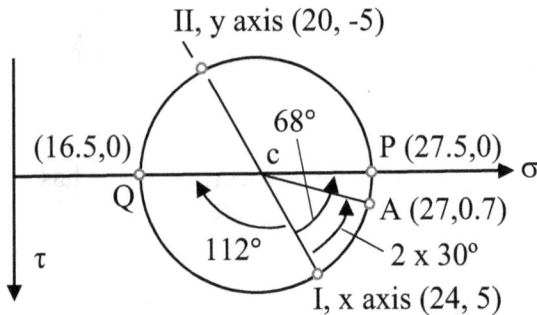

II, y axis (20, -5)

(16.5,0)

68°

c

P (27.5,0)

σ

A (27,0.7)

112°

2 x 30°

τ

I, x axis (24, 5)

a) - measure $2\theta = 2 \times 30°$ counter-clockwise from point I and
    mark Point A.
   - measure the coordinates of A to find the stresses:
    $\sigma = 27$ MPa, $\tau = 0.7$ MPa.

b). The principal stresses are at the right most and left most
   points (P and Q):
   at point P, $\sigma_1 = 27.5$ MPa,
   at point Q, $\sigma_2 = 16.5$ MPa.

c). The principal directions are measured directly from Mohr's
   circle.

P is 68° from Point I (x axis) counter-clockwise and Q is 112° from Point I (x axis) clockwise. Therefore,

$\theta_1 = 68°/2 = 34°$ from x axis,
$\theta_2 = -112°/2 = -56°$ from x axis.

The principal directions are shown on the right.

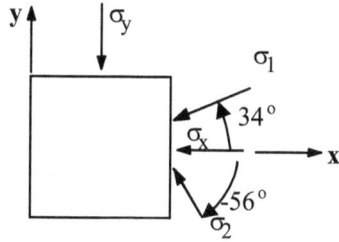

4.4 Polar Coordinates and Stress Transformation

Stress tensor

In earlier discussions, it was indicated that the stresses in two dimensions can be represented by either the three stress components in a Cartesian coordinate system $\{\sigma_x, \sigma_y, \tau_{xy}\}$ or the principal components $\{\sigma_1, \sigma_2\}$ plus their directions. The directions of x and y are chosen arbitrarily and the values of the three stresses $\{\sigma_x, \sigma_y, \tau_{xy}\}$ vary with the direction relative to that of σ_1. However, the principal stresses will remain the same in both magnitude and direction no matter what x and y directions are used. Using a similar concept of a vector in one dimension, the stresses in two dimensions are a stress tensor.

A stress tensor is best defined by its principal components and directions. The stress components in any of the other directions will have different values. For example, the normal stress in a stress tensor can be considered as an ellipse with σ_1 and σ_2 as the longer and shorter semi axes, respectively, as shown in Fig. 4.6.

The x, y coordinate system is defined by an angle θ from the x axis with respect to the major principal direction (i.e., the σ_1 direction). θ can vary from 0° to 360° from the σ_1 direction. The normal and shear stresses (σ_x and τ_{xy}) in the x direction can be directly calculated by Eqns. 4.11 and 4.12. The stresses in the y direction (σ_y and τ_{xy}) can also be calculated by these two equations if θ is replaced by ($\theta \pm 90°$). Therefore, for the same stress tensor, there are infinite sets of $\{\sigma_x, \sigma_y, \tau_{xy}\}$ as the angle θ changes.

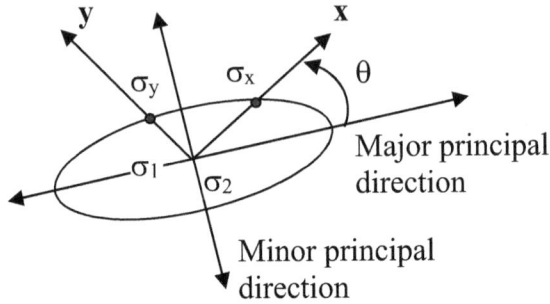

Fig. 4.6 Normal stress variation of a stress tensor.

Cylindrical polar coordinate system

Stresses around an underground opening are very complex. Although the Cartesian coordinate system x, y is commonly used, it will be simpler in analysis if a Polar coordinate system is introduced to coincide with the shape of a circular opening. The Polar coordinate system is defined by two new coordinates r and θ, as shown in Fig. 4.7. The coordinates of point P can be described by either (x, y) or (r, θ). Transformation between the two sets of coordinates is given by

$$\begin{cases} r = \sqrt{x^2 + y^2} \\ \theta = \text{Arc Tan } (y/x) \end{cases} \tag{4.16a}$$

and $\begin{cases} x = r \cos \theta \\ y = r \sin \theta \end{cases}$ $\tag{4.16b}$

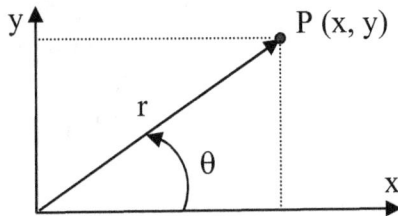

Fig. 4.7 Polar and xy coordinates.

Stress transformation

If the stress components in one coordinate system, such as in the x, y system, $\{\sigma_x,\ \sigma_y,\ \tau_{xy}\}$ are known, the stress components in another coordinate system (r, θ system), $\{\sigma_r,\ \sigma_\theta,\ \tau_{r\theta}\}$ can be calculated through stress transformation.

Suppose that the stress components at the same location defined by (x, y) or (r, θ) in the two systems are as illustrated in Fig. 4.8. Since the angle is θ measured counter-clockwise from the x axis, Eqns. 4.5 and 4.7 can be used directly to calculate the stresses $\{\sigma_r,\ \tau_{r\theta}\}$ in the r direction.

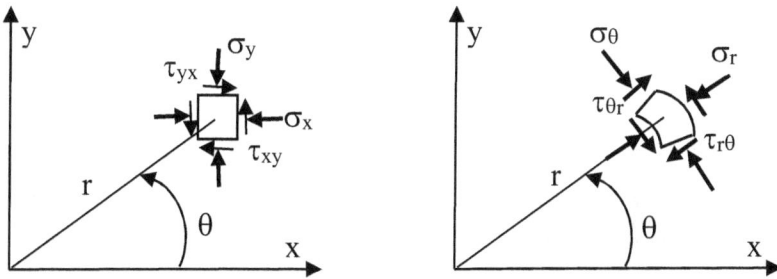

Fig. 4.8 stress transformation.

$$\sigma_r = \frac{(\sigma_x+\sigma_y)}{2} + \frac{(\sigma_x-\sigma_y)}{2} \cos 2\theta + \tau_{xy} \sin 2\theta$$

and $$\tau_{r\theta} = -\frac{(\sigma_x-\sigma_y)}{2} \sin 2\theta + \tau_{xy} \cos 2\theta$$

To calculate the stresses $\{\sigma_\theta,\ \tau_{\theta r}\}$ in the θ direction at the same point, replace θ by (θ+90°) in Eqn. 4.5. We have

$$\sigma_\theta = \frac{(\sigma_x+\sigma_y)}{2} - \frac{(\sigma_x-\sigma_y)}{2} \cos 2\theta - \tau_{xy} \sin 2\theta$$

and $\tau_{\theta r} = -\tau_{r\theta}$

If the above three equations are re-arranged to separate the three stress components (σ_x, σ_y and τ_{xy}), they can be written in a matrix form as follows:

$$\begin{Bmatrix} \sigma_r \\ \sigma_\theta \\ \tau_{r\theta} \end{Bmatrix} = \begin{bmatrix} \dfrac{1+\cos 2\theta}{2} & \dfrac{1-\cos 2\theta}{2} & \sin 2\theta \\ \dfrac{1-\cos 2\theta}{2} & \dfrac{1+\cos 2\theta}{2} & -\sin 2\theta \\ -\dfrac{\sin 2\theta}{2} & \dfrac{\sin 2\theta}{2} & \cos 2\theta \end{bmatrix} \begin{Bmatrix} \sigma_x \\ \sigma_y \\ \tau_{xy} \end{Bmatrix} \quad (4.17)$$

Equation 4.17 is for stress transformation from the x, y system to the r, θ system.

Similarly, stress transformation from the r, θ system to the x, y system can be derived by modifying Eqn. 4.17: simply replacing θ by -θ. In this case, the x axis is rotated an angle θ from the r axis clockwise, which means negative in Eqns. 4.5 and 4.7. All terms with sin2θ become negative and all terms with cos2θ are unchanged. The transformation is given in Eqn. 4.18.

$$\begin{Bmatrix} \sigma_x \\ \sigma_y \\ \tau_{xy} \end{Bmatrix} = \begin{bmatrix} \dfrac{1+\cos 2\theta}{2} & \dfrac{1-\cos 2\theta}{2} & -\sin 2\theta \\ \dfrac{1-\cos 2\theta}{2} & \dfrac{1+\cos 2\theta}{2} & \sin 2\theta \\ \dfrac{\sin 2\theta}{2} & -\dfrac{\sin 2\theta}{2} & \cos 2\theta \end{bmatrix} \begin{Bmatrix} \sigma_r \\ \sigma_\theta \\ \tau_{r\theta} \end{Bmatrix} \quad (4.18)$$

Alternatively, it can also be derived by inversion of Eqn. 4.17.

Example 4.6

Given the stresses in the x, y coordinate system as $\sigma_x = 24$ MPa, $\sigma_y = 20$ MPa and $\tau_{xy} = 5$ MPa. Determine the stresses in another coordinate system: x', y' with the x' axis at 30° from the x axis counter-clockwise.

Solutions:

This question is the same as the transformation condition in Eqn. 4.17 with θ = 30°. First, calculate the following terms:

$\cos 2\theta = \cos (2 \times 30°) = 0.5$, $\sin 2\theta = \sin (2 \times 30°) = 0.866$
$(1 + \cos2\theta)/2 = 0.75$, $(1 - \cos 2\theta)/2 = 0.25$

Then substitute the above values into Eqn. 4.17,

$$\begin{Bmatrix} \sigma_{x'} \\ \sigma_{y'} \\ \tau_{x'y'} \end{Bmatrix} = \begin{bmatrix} 0.75 & 0.25 & 0.866 \\ 0.25 & 0.75 & -0.866 \\ -0.433 & 0.433 & 0.5 \end{bmatrix} \begin{Bmatrix} 24 \\ 20 \\ 5 \end{Bmatrix} = \begin{Bmatrix} 27.3 \\ 16.7 \\ 0.77 \end{Bmatrix}$$

$\therefore \sigma_{x'} = 27.3$ MPa, $\sigma_{y'} = 16.7$ MPa and $\tau_{x'y'} = 0.77$ MPa.

4.5 Concept of Three-dimensional Stresses

In the field, the stresses are in a three-dimensional state. To fully define the state of stress at a point within a rock mass, it is necessary to consider an infinitesimal cubic volume enclosing the point in question, with each pair of faces perpendicular to the three axes, x, y and z, respectively, as shown in Fig. 4.9 (where only three faces are shown and the three opposite faces are hidden). In this case, on each face there are three stress components: one normal stress and two shear stresses. The stresses on the three opposite faces are equal but pointing in opposite directions to keep the cube in equilibrium. Therefore, there are a total of nine stress components.

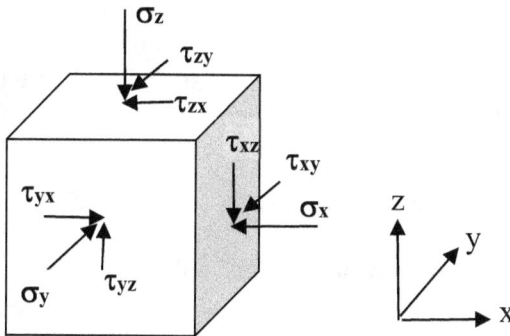

Fig. 4.9 Stresses in three dimensions.

However, each of the three pairs of shear stresses has equal components,

i.e., $\tau_{xy} = \tau_{yx}, \tau_{yz} = \tau_{zy}, \tau_{xz} = \tau_{zx}$ (4.19) and they are called conjugate stresses. That leaves six independent stress components,

$$\{\sigma\} = \{\sigma_x, \sigma_y, \sigma_z, \tau_{xy}, \tau_{yz}, \tau_{xz}\} \quad (4.20)$$

Similar to the case in 2D, for a given stress state, the value of each stress component in Eqn. 4.20 varies with the coordinate axes directions. However, the principal stresses remain unchanged. In a three-dimensional stress field, there are three principal stresses, which can be determined by solving the following equation for the roots of σ.

$$\sigma^3 + I_1 \sigma^2 + I_2 \sigma + I_2 = 0 \quad (4.21)$$

where I_1, I_2 and I_3 are called stress invariants, given by

$$\begin{cases} I_1 = -(\sigma_x + \sigma_y + \sigma_z) \\ I_2 = \sigma_x \sigma_y + \sigma_y \sigma_z + \sigma_x \sigma_z - \tau_{xy}^2 - \tau_{yz}^2 - \tau_{xz}^2) \\ I_3 = -(\sigma_x \sigma_y \sigma_z - \sigma_x \tau_{yz}^2 - \sigma_y \tau_{xz}^2 - \sigma_z \tau_{xy}^2 + 2\tau_{xy}\tau_{yz}\tau_{xz}) \end{cases} \quad (4.22)$$

The principal stresses are defined as

$$\sigma_1 \geq \sigma_2 \geq \sigma_3 \quad (4.23)$$

In the principal directions, the shear stresses are zero. More detail in determination of stresses on a plane in three dimensions has been given in other books (such as Jaeger and Cook 1976). The detail is beyond the scope of this book.

4.6 Stress Strain Relationships

It is understood that an admissible solution to any problem in solid mechanics must satisfy both the differential equations of static equilibrium and the equations of strain compatibility. The way in which stress and strain are related in a material under load can be described by its constitutive behaviour. A variety of constitutive models have been formulated for various engineering materials which describe both the time-independent and time-dependent responses of the materials to the applied load. These models describe responses in terms of elasticity, plasticity, viscosity and creep, as well as combinations of these modes. For any constitutive model, stress and strain, or some derived quantities, such as stress and strain rates, are related through a set of constitutive equations. Elasticity represents the most common constitutive behaviour of engineering materials including many rocks, and it forms a useful basis for description of more complex behaviour.

In Chapter 2, the simplest model of one-dimensional loading was discussed. For a simplified linear elastic model, the stress-strain relationship was given in Eqn. 2.1 as a linear equation. In the field however there are six independent stress components and correspondingly there are six independent strain components. In this case, a constitutive model would relate the six stresses with the six strains. These relationships can be given as six linear equations for linear elastic materials and be expressed in a matrix form. The stresses are defined as a stress vector $\{\sigma\}$ and the strains as a strain vector $\{\varepsilon\}$:

$$\{\sigma\} = \begin{Bmatrix} \sigma_x \\ \sigma_y \\ \sigma_z \\ \tau_{xy} \\ \tau_{yz} \\ \tau_{zx} \end{Bmatrix}, \text{ and } \{\varepsilon\} = \begin{Bmatrix} \varepsilon_x \\ \varepsilon_y \\ \varepsilon_z \\ \gamma_{xy} \\ \gamma_{yz} \\ \gamma_{zx} \end{Bmatrix} \qquad (4.24)$$

The most general statement of linear elastic constitutive behaviour is a generalized form of Hook's Law, in which any strain component is a linear function of all the stress components, i.e.,

$$\begin{Bmatrix} \varepsilon_x \\ \varepsilon_y \\ \varepsilon_z \\ \gamma_{xy} \\ \gamma_{yz} \\ \gamma_{zx} \end{Bmatrix} = \begin{bmatrix} S_{11} & S_{12} & S_{13} & S_{14} & S_{15} & S_{16} \\ S_{21} & S_{22} & S_{23} & S_{24} & S_{25} & S_{26} \\ S_{31} & S_{32} & S_{33} & S_{34} & S_{35} & S_{36} \\ S_{41} & S_{42} & S_{43} & S_{44} & S_{45} & S_{46} \\ S_{51} & S_{52} & S_{53} & S_{54} & S_{55} & S_{56} \\ S_{61} & S_{62} & S_{63} & S_{64} & S_{65} & S_{66} \end{bmatrix} \begin{Bmatrix} \sigma_x \\ \sigma_y \\ \sigma_z \\ \tau_{xy} \\ \tau_{yz} \\ \tau_{zx} \end{Bmatrix} \qquad (4.25)$$

or, $\{\varepsilon\} = [S]\{\sigma\}$ \hfill (4.25a)

Each of the elements S_{ij} of the matrix [S] is called a compliance or an *elastic modulus*. Although there are 36 moduli in Eqn. 4.25, it can be demonstrated that the compliance matrix is symmetric and therefore there are only 21 independent constants in matrix [S].

In some cases, it is more convenient to apply Eqn. (4.25a) in an inversed form, i.e.,

$$\{\sigma\} = [D] \{\varepsilon\} \tag{4.26}$$

The matrix [D] is the inverse of [S], called the *elasticity matrix* or the matrix of *elastic stiffness*. For general anisotropic elasticity there are 21 independent values of stiffness. In the case of isotropic elastic materials, Eqn. 4.25, reduces to a simplified form shown in Eqn. 4.27. In this matrix, most elements are zero and the rest along the diagonal line are composed of E and v, which we are already familiar with.

$$
\begin{Bmatrix} \varepsilon_x \\ \varepsilon_y \\ \varepsilon_z \\ \gamma_{xy} \\ \gamma_{yz} \\ \gamma_{zx} \end{Bmatrix} = \frac{1}{E}
\begin{bmatrix}
1 & -\upsilon & -\upsilon & 0 & 0 & 0 \\
-\upsilon & 1 & -\upsilon & 0 & 0 & 0 \\
-\upsilon & -\upsilon & 1 & 0 & 0 & 0 \\
0 & 0 & 0 & 2(1+\upsilon) & 0 & 0 \\
0 & 0 & 0 & 0 & 2(1+\upsilon) & 0 \\
0 & 0 & 0 & 0 & 0 & 2(1+\upsilon)
\end{bmatrix}
\begin{Bmatrix} \sigma_x \\ \sigma_y \\ \sigma_z \\ \tau_{xy} \\ \tau_{yz} \\ \tau_{zx} \end{Bmatrix} \tag{4.27}
$$

The common equations of Hook's law for isotropic elasticity are readily derived from Eqn. 4.27,

e.g.,
$$
\begin{cases}
\varepsilon_x = \frac{1}{E}[\sigma_x - \upsilon(\sigma_y + \sigma_z)], \text{etc.} \\
\gamma_{xy} = \frac{1}{G}\tau_{xy}, \text{etc.}
\end{cases} \tag{4.28}
$$

where $G = E /2(1+\upsilon)$ \hfill (4.29)

G is called the shear modulus. The other stress or strain components can be defined in the same way by replacing the subscripts x, y, z in Eqn. 4.28 in sequence.

For isotropic materials, determination of any two constants out of the three (E, G and υ) will give a complete set of characteristics of the material. This can be accomplished by uniaxial compressive tests as described earlier.

Often, the inversed form of the stress-strain equation in Eqn. 4.27 for isotropic materials is used as in the following:

$$\begin{Bmatrix} \sigma_x \\ \sigma_y \\ \sigma_z \\ \tau_{xy} \\ \tau_{yz} \\ \tau_{zx} \end{Bmatrix} = \frac{G}{(1-2\upsilon)} \begin{bmatrix} 2(1-\upsilon) & 2\upsilon & 2\upsilon & 0 & 0 & 0 \\ 2\upsilon & 2(1-\upsilon) & 2\upsilon & 0 & 0 & 0 \\ 2\upsilon & 2\upsilon & 2(1-\upsilon) & 0 & 0 & 0 \\ 0 & 0 & 0 & (1-2\upsilon) & 0 & 0 \\ 0 & 0 & 0 & 0 & (1-2\upsilon) & 0 \\ 0 & 0 & 0 & 0 & 0 & (1-2\upsilon) \end{bmatrix} \begin{Bmatrix} \varepsilon_x \\ \varepsilon_y \\ \varepsilon_z \\ \gamma_{xy} \\ \gamma_{yz} \\ \gamma_{zx} \end{Bmatrix} \quad (4.30)$$

The commonly used explicit equations of Eqn. 4.30 are given as

$$\begin{cases} \sigma_x = \frac{2G}{(1-2\upsilon)}[(1-\upsilon)\varepsilon_x + \upsilon(\varepsilon_y + \varepsilon_z)], \text{etc.} \\ \tau_{xy} = G\,\gamma_{xy}, \text{etc.} \end{cases} \quad (4.31)$$

Plane Strain Condition

For a drift excavated in the field with its axis parallel to the y axis, the deformation pattern in any cross section of the drift should be the same within the x-z plane (Fig. 4.10) except for near the ends of the drift. This is a plane strain condition, where

$\varepsilon_y = 0, \gamma_{xy} = \gamma_{yz} = 0.$

From Eqn. 4.28, let $\varepsilon_y = 0$, we have $E\,\varepsilon_y = \sigma_y - \upsilon\,(\sigma_x + \sigma_z) = 0$.

Then, $\sigma_y = \upsilon\,(\sigma_x + \sigma_z)$ \hfill (4.32)

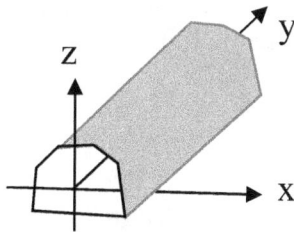

Fig. 4.10 Plane strain condition in cross sections of a long drift.

It means that under the plane strain condition, the stress along the drift axis is not zero, although there is no deformation in that direction. Furthermore, the stress-strain relationships for the plane strain condition can be derived from Eqns. 4.28 and 4.31, with consideration of Eqn. 4.32:

$$\begin{cases} \varepsilon_x = \dfrac{1\text{-}\upsilon^2}{E}\left(\sigma_x - \dfrac{\upsilon}{1\text{-}\upsilon}\,\sigma_z\right) \\[2mm] \varepsilon_z = \dfrac{1\text{-}\upsilon^2}{E}\left(\sigma_z - \dfrac{\upsilon}{1\text{-}\upsilon}\,\sigma_x\right) \\[2mm] \gamma_{xz} = \dfrac{2(1+\upsilon)}{E}\,\tau_{xz} \end{cases} \tag{4.33}$$

and
$$\begin{cases} \sigma_x = \dfrac{E}{(1+\upsilon)(1-2\upsilon)}\big((1-\upsilon)\varepsilon_x + \upsilon\,\varepsilon_z\big) \\[2mm] \sigma_z = \dfrac{E}{(1+\upsilon)(1-2\upsilon)}\big((1-\upsilon)\varepsilon_z + \upsilon\,\varepsilon_x\big) \\[2mm] \tau_{xz} = \dfrac{E}{2(1+\upsilon)}\,\gamma_{xz} \end{cases} \tag{4.34}$$

Transverse Isotropic Rocks

Some rocks, such as stratified rocks and rocks with strong bedding planes, have strong directional effect. They normally show isotropic characteristics within a rock layer (the x-y plane) but are very different in the 3rd direction (z). This feature is referred to as transverse isotropic.

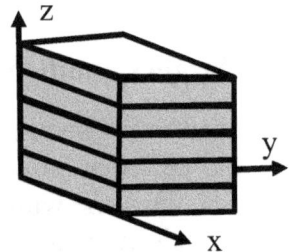

For transversely isotropic elastic materials, there are two sets of material constants: E_1 and υ_1, within the isotropic plane, and E_2 and υ_2 in the z direction perpendicular to the plane. The stress-strain relationship defined in Eqn. 4.27 is modified as below. More detail can be found in (Brady and Brown 2006).

$$\begin{Bmatrix} \varepsilon_x \\ \varepsilon_y \\ \varepsilon_z \\ \gamma_{xy} \\ \gamma_{yz} \\ \gamma_{zx} \end{Bmatrix} = \frac{1}{E_1} \begin{bmatrix} 1 & -\upsilon_1 & -\upsilon_2 & 0 & 0 & 0 \\ -\upsilon_1 & 1 & -\upsilon_2 & 0 & 0 & 0 \\ -\upsilon_2 & -\upsilon_2 & E_1/E_2 & 0 & 0 & 0 \\ 0 & 0 & 0 & 2(1+\upsilon_1) & 0 & 0 \\ 0 & 0 & 0 & 0 & E_1/G_2 & 0 \\ 0 & 0 & 0 & 0 & 0 & E_1/G_2 \end{bmatrix} \begin{Bmatrix} \sigma_x \\ \sigma_y \\ \sigma_z \\ \tau_{xy} \\ \tau_{yz} \\ \tau_{zx} \end{Bmatrix} \tag{4.35}$$

where $G_2 = E_2/2(1+\upsilon_2)$ $\tag{4.36}$

Chapter 5

Strength of Rock and Failure Criteria

Rock like any other materials will fail once the stress generated by external load is beyond a critical level. The maximum allowable stress, as discussed before, is the strength of the material. An important question is: how could the failure take place? Different types of rock may fail in different ways when the loading conditions and environment change. Some fails gradually and quietly and others abruptly and violently.

As shown in the margin sketch, a cylindrical specimen may fail in the axial direction splitting parallel to the direction of uniaxial compression. Shear failure also occurs frequently in uniaxial compression. The shear plane forms an acute angle with the direction of compression. In triaxial tests, shear failures invariably occur and two slip surfaces or shear fracture planes will develop at an inclined angle less than 45° to the axial stress direction.

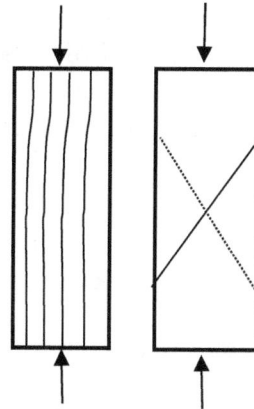

Under uniaxial tension, however, specimens almost exclusively fail on weak planes, which are most vulnerable when they are nearly perpendicular to the direction of the tensile loading. Fracture theory developed in continuum mechanics has also been applied to explain the fracture phenomena in rocks. It works well for brittle materials under tension but may not be so satisfactory for rocks under high compression.

In the following sections, only those failure criteria that have been used commonly in rock engineering will be discussed. They include the maximum tensile and compressive stress criteria, Coulomb, Mohr's, and Empirical criterion.

The failure mode also depends on the failure mechanism. In engineering, we are not just interested in determining when the material is going to fail; it is more important for us to avoid the failure and to estimate the maximum possible load before failure. To make an engineering judgement, we need some criterion, which varies with different failure mechanisms.

5.1 Maximum Tension and Compression

If the failure is simply due to tension or compression, the failure criterion is quite simple. Under uniaxial tension, a rock sample fails once the tensile stress reaches its tensile strength:

$$|\sigma_3| \geq \sigma_t \tag{5.1}$$

Note: The above criterion only considers stress magnitude. If $\sigma_3 < 0$, σ_3 is a tensile stress. If $\sigma_3 \geq 0$, there is no tensile failure.

Examples of rock failure in tension include failure in the Brazilian test, as discussed before, and pillar spalling observed underground (Fig. 5.1).

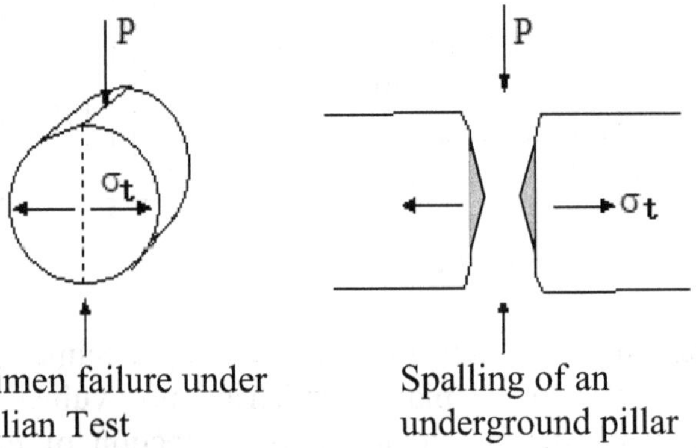

Specimen failure under Brazilian Test   Spalling of an underground pillar

Fig. 5.1 Examples of tensile failure of rock.

Under uniaxial compression, a rock sample fails once the compressive stress reaches its compressive strength as follows:

$$\sigma_1 \geq \sigma_c \tag{5.2}$$

5.2 Coulomb Criterion --- for Shear Failure

With this criterion, shear failure occurs once the induced shear stress, regardless of its sign, on a plane (Fig. 5.2) reaches the shear strength of that plane (-τ only indicates shearing in the opposite direction).

$$|\tau| \geq \tau_s \qquad (5.3)$$

where $\tau_s = c + \sigma\mu$, μ is the coefficient of friction, given by

$$\mu = \tan\phi$$

and ϕ is the friction angle of the rock surface. This equation has the same format as the shear failure on a surface as discussed before.

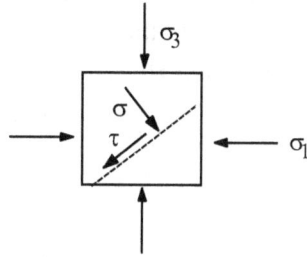

Fig. 5.2 Stresses on a plane.

It is important to understand that shear failure may also occur even not under direct shear loading. Shear failure becomes possible provided that the induced shear stress exceeds the shear strength on a plane even if the material is in compression overall. Under compression, the shear stress on a plane can be calculated by Eqn. 4.12 and the corresponding failure criterion becomes

$$|\tau| = \left|\frac{\sigma_1 - \sigma_3}{2} \sin 2\theta\right| \geq c + \mu\sigma \qquad (5.4)$$

It is also important to know that shear failure may not necessarily occur on the plane where the maximum shear stress occurs because the shear strength τ_s could be lower on other planes, particularly when there is a weak discontinuity. This is especially important for rock masses which contain numerous weak discontinuities in random directions.

Example 5.1

Given: In a compressive stress field, $\sigma_1 = 27.4$ MPa, $\sigma_3 = 16.6$ MPa, with σ_1 in the x direction. Assume that a rock mass has an internal frictional angle $\phi = 25°$ and cohesion $c = 0$.

a) Is shear failure possible on the maximum shear plane?

b) Is shear failure possible on a joint plane whose normal is 30° from σ_1 with a friction angle $\phi = 18°$?

Solutions:

a). From discussions in the previous chapter, the maximum shear stress occurs on the plane, the normal of which is $\theta = \pm 45°$ (two possible planes) from the σ_1 direction and the stresses on the maximum shear planes (Eqn. 4.13) are:

$$\sigma = (27.4 + 16.6)/2 = 22 \text{ MPa, and}$$
$$|\tau| = (27.4 - 16.6)/2 = 5.4 \text{ MPa.}$$

By the criterion in Eqn. 5.3, the shear strength on the plane is

$$\tau_s = c + \sigma \tan \phi = 0 + 22 \times \tan 25° = 10.3 \text{ MPa.}$$

Because $|\tau| = 5.4$ MPa $< \tau_s$, no shear failure would take place on the maximum shear planes. However, we may need to consider a safe margin when making a judgement.

b). The direction of the joint plane is defined by its normal. Hence $\theta = \pm 30°$ (two possible planes on both sides of σ_1). By Eqns. 4.11 and 4.12 the stresses on the joint plane are

$$\sigma = \frac{27.4 + 16.6}{2} + \frac{27.4 - 16.6}{2} \cos(2 \times \pm30°)$$
$$= 22 + 5.4 \times 0.5 = 24.7 \text{ MPa.}$$

$$\tau = -\frac{27.4 - 16.6}{2} \sin(2 \times \pm30°)$$
$$= \pm 5.4 \times 0.866 = \pm4.7 \text{ MPa.}$$

The shear strength on the joint plane is

$$\tau_s = \sigma \tan \phi = 24.7 \times \tan 18° = 8.0 \text{ MPa.}$$

Since $|\tau| = 4.7$ MPa $< \tau_s$, no shear failure is possible on the joint.

5.3 Mohr's Criterion --- for Shear Failure

Based on Mohr's Criterion, there exists a functional relationship between τ and τ_s for the material. Shear failure takes place on a plane when

$$|\tau| \geq f(\sigma) \tag{5.5}$$

In the previous Coulomb criterion, this relationship f was assumed to be a linear function of σ (Eqn. 5.3). This however may not be the same for every case and the relationship needs to be established by laboratory tests and Mohr's circle analysis.

The following describes the procedure to develop the failure strength envelope:
- conduct a series of triaxial compressive tests on rock samples at various confining pressure (σ_3).
- record the values of σ_1 and σ_3 at failure for each test,
- draw a Mohr circle for each test based on σ_1 and σ_3 (Note: in uniaxial compressive test, $\sigma_3 = 0$),
- draw a line or curve tangent to all circles. This is the Mohr's strength envelope as shown in Fig. 5.3.

A simplified Mohr's envelope is represented by a straight line

$$\tau_s = c_i + \sigma \tan \phi_i \tag{5.6}$$

where the frictional angle ϕ_i and the corresponding cohesion c_i may not be constant and would vary along the envelope.

Any state of stress which generates a Mohr circle beyond the Mohr's envelope will cause instability and failure will occur in

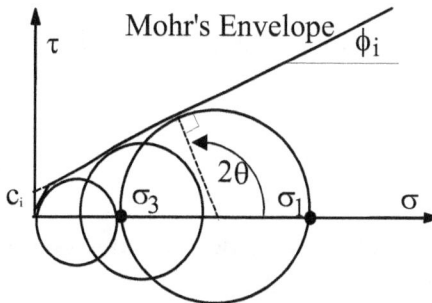

Fig. 5.3 Mohr's criterion.

the direction of θ with respect to σ_1 as shown in Fig. 5.3. The direction of 2θ on the Mohr diagram is determined by a radius perpendicular to the failure envelope at the intersection point. It can be shown in Fig. 5.3, $2\theta = 90° + \phi_i$,

$$\therefore \qquad \theta = 45° + \phi_i/2 \qquad (5.7a)$$

This is the failure direction - the normal of the failure plane relative to the maximum principal stress σ_1, shown in Fig. 5.4.

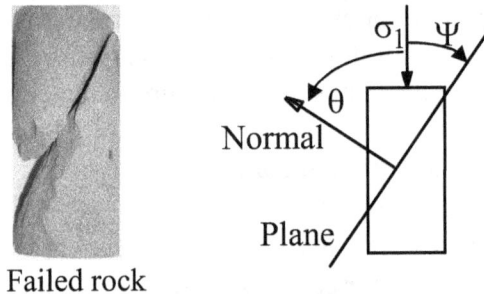

Failed rock

Fig. 5.4 Failure angle.

Since $\theta + \Psi = 90°$, the angle of the failure plane itself is determined as

$$\Psi = 45° - \phi_i/2 \qquad (5.7b)$$

Both angles θ and Ψ are measured from σ_1. Equation 5.7b also implies that the internal friction angle can be estimated from the failure angle of the failed rock sample as shown in Fig. 5.4.

$$\phi_i = 2 (45° - \Psi) \qquad (5.7c)$$

If the rock has a weak plane, failure possibility should be checked on both the weak plane and the plane defined by Eqn. 5.7b.

Example 5.2

A number of triaxial tests have been conducted on a type of rock samples. The results generated a simplified Mohr envelope which can be represented by

$$\tau = 10 \text{ (MPa)} + \sigma \tan 41°.$$

A pillar is designed in the rock and the estimated vertical stress is 90 MPa. Assess the pillar stability by Mohr's criterion.

Solutions:

- draw the Mohr envelope: $\tau = 10$ (MPa) $+ \sigma \tan 41°$ in the σ - τ plane as shown in Fig. 5.5,
- Draw a Mohr circle through the two points on the σ axis at $\sigma_3 = 0$ MPa and $\sigma_1 = 90$ MPa.

Then make a judgement based on the graph: If the circle touches or goes above the strength envelope, the pillar is unstable; if it is below the envelope, the pillar is stable. From Fig. 5.5, the pillar seems stable. In practice, however, we may need to consider a factor of safety.

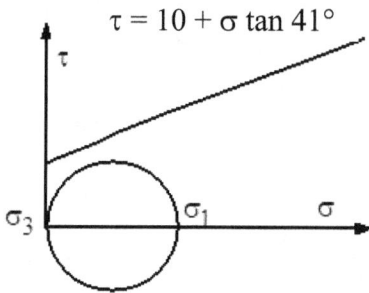

$$\tau = 10 + \sigma \tan 41°$$

Fig. 5.5 Mohr envelop solution.

5.4 Griffith Theory - Tensile Failure at the Crack Tips

Griffith theory is developed in fracture mechanics. The general assumptions are that a) any material contains numerous randomly oriented thin elliptical cracks, and b) all cracks remain open. As the external stress reaches some level, an extremely high tensile stress concentration occurs at the tips of the cracks. Failure occurs when the tensile stress concentration exceeds the tensile strength,

$$\begin{cases} (\sigma_1 - \sigma_3)^2 = 8 \sigma_t(\sigma_1 + \sigma_3), & \text{if } \sigma_1 + 3\sigma_3 > 0 \\ \sigma_3 = -\sigma_t, & \text{if } \sigma_1 + 3\sigma_3 < 0 \end{cases} \tag{5.8}$$

Since rock mass is under compression all the time in the field except near the opening surfaces, assumption b) does not always represent the reality in rock engineering. Therefore, Griffith theory, as is, has limited application. Some researchers have been trying to modify this theory for application in rock engineering (see Jaeger and Cook 1976). However, from a practical point of view, direct use of this failure criterion in rock engineering seems to have some difficulty and a more straight-forward and practical approach will be more useful. The most commonly used approach is the Hoek and Brown empirical failure criterion, which will be described in more detail below.

5.5 Empirical Failure Criterion for Jointed Rock Masses

Rock mass is a very complex system. It consists of different types of rocks, joints and other weak and randomly oriented discontinuities. The behaviour of a rock mass generally cannot be described by one simple model. The criteria discussed above may apply to simplified cases but to the rock masses in the field, quite often, the results can differ significantly from reality. For example, Hoek and Brown (1980a) demonstrated that at high stress level, both the original and the modified Griffith theories do not match the results of laboratory tests. The original theory is too conservative (presumably because of assumption b - unrealistic as mentioned earlier) and the modified theory is too risky.

Rather than developing a difficult mathematics model, Hoek and Brown (1980b) proposed an empirical failure criterion for the rock masses encountered in the field, using results from laboratory tests of numerous rock samples. A conceptual model of the empirical failure criterion is illustrated in Fig. 5.6.

This criterion was been updated and refined later by the authors (Hoek and Brown 1988). Readers are strongly encouraged to read the latest updates on this method as it has been continuously updated by many practical professionals as more field data become available.

When the criterion was developed, the following objectives were set: a) it should agree with experimental data, b) it should be expressed by a simple equation, and c) it can be extended to cover anisotropic failure and the failure of jointed rock masses. The criterion was initially derived for applications in underground excavation design and it was therefore expressed in terms of the effective major and minor principal stresses in the rock mass.

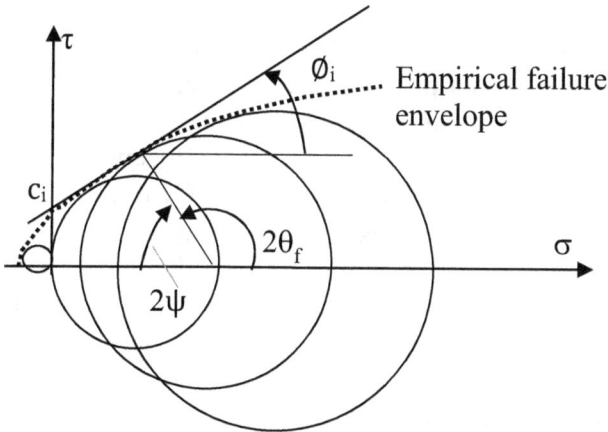

Fig. 5.6 Illustration of a strength envelope compatible with test results.

This empirical failure criterion looks very similar to the Mohr's failure criterion discussed earlier. A comparison of Figs. 5.3 and 5.6 will show the similarities. An improvement in the empirical criterion over the Mohr's criterion is that empirical constants are introduced to account for many types of rock mass conditions encountered in the field (not just rock samples in laboratory scale).

Another important feature is that water pressure, **p**, on joints is taken into consideration. As mentioned before, the water pressure effectively reduces the normal stress on a surface by an amount equal to its magnitude. The resultant normal stress is called the effective stress, given by

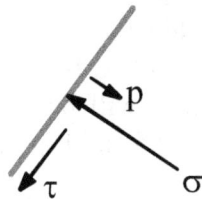

$$\sigma' = \sigma - p \qquad\qquad (5.9)$$

In the following, normal stresses, σ_1', etc., are all expressed in effective stress values.

1) Compressive strength of jointed rock masses

The basic equation defining the empirical failure criterion is intended to match the stress state at failure in terms of principal stresses with consideration of field variations. The general form for the strength of a rock mass is given by

$$\sigma_1' = \sigma_3' + (m\,\sigma_c\,\sigma_3' + s\,\sigma_c^2)^a \qquad (5.10)$$

where σ_1' is the major principal effective stress at failure,

σ_3' is the minor principal effective stress at failure,

a, m and s are the empirical material constants, depending on the quality of the rock mass. In general,

m > 0, with values ranging from 0.001 to 33,

s > 0, with values ranging from 0.0 to 1.0.

The reliability of using Eqn. 5.10 to predict the strength of a rock mass, to a great extent, depends on how accurate the selected values of these constants are. If appropriate values are chosen, Eqn. 5.10 can be applied to intact rock and rock masses which have good or poor quality. The following is a guide-line in selecting those constants.

Intact rock: For rock specimens and rocks free of weakness or defects, a = 0.5 and s = 1.0. Equation 5.10 becomes

$$\sigma_1' = \sigma_3' + \sqrt{m_i\,\sigma_c\,\sigma_3' + \sigma_c^2} \qquad (5.10a)$$

where m_i represents the highest m value of the intact rock and varies with the type of rock.

Rock masses with good quality: For rock masses in the field which have good to reasonable quality and their strength is controlled by tightly interlocking angular rock pieces (further detail will be given in a later chapter on rock mass quality), a = 0.5 and Eqn. 5.10 becomes

$$\sigma_1' = \sigma_3' + \sqrt{m_b\,\sigma_c\,\sigma_3' + s\,\sigma_c^2} \qquad (5.10b)$$

where m_b is the constant for rock masses, related to m_i, to be discussed more later.

Rock masses with poor quality: For the rock mass conditions where the interlocking has been partially destroyed by shearing or weathering and the rock mass has no tensile strength or "cohesion", a modified criterion (Hoek et al. 1995) is to be used. In this case, $s = 0$ and Eqn. 5.10 is reduced to

$$\sigma_1' = \sigma_3' + \sigma_c \, (m_b \frac{\sigma_3'}{\sigma_c})^a \qquad (5.10c)$$

In accordance to different loading conditions, the above set of equations for jointed rock masses are simplified to the following forms:

Under uniaxial compression, $\sigma_3' = 0$. Equation 5.10b will give the uniaxial compressive strength of the rock mass,

$$\sigma_1' \,(= \sigma_{c\text{-mass}}) = \sqrt{s}\,\sigma_c \leq \sigma_c \qquad (5.11)$$

Under uniaxial tension, $\sigma_1' = 0$. Re-arrange Eqn. 5.10b and take the smallest root for σ_3', it gives the uniaxial tensile strength of the rock mass,

$$\therefore \; \sigma_3' \,(= \sigma_{t\text{-mass}}) = \frac{\sigma_c}{2}\,(m_b - \sqrt{m_b^2 + 4\,s}) \qquad (5.12)$$

The above loading conditions corresponding to the failure envelope are demonstrated in Fig. 5.7.

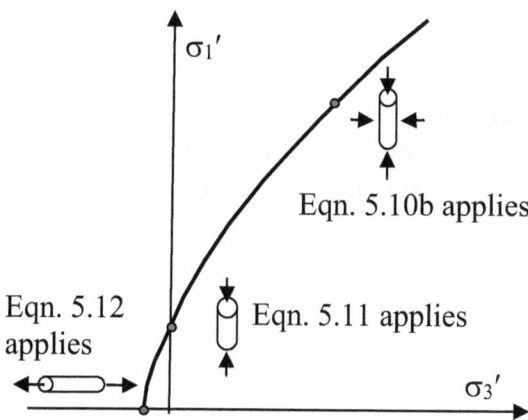

Fig. 5.7 Illustrated empirical failure envelope for different loading conditions.

In applying the empirical failure criterion in underground design, the following data are needed to determine the rock mass strength by Eqn. 5.10: the major and minor principal stresses, σ_1 and σ_3, in the field (to be measured or estimated otherwise) and the water pressure. The calculated strength σ_1' is then compared with the effective stress in the field σ_1 to determine the stability. If σ_1 is less than the calculated strength with consideration of a safety factor (SF), i.e.,

$$\sigma_1 < \sigma_1' / SF \qquad (5.12a)$$

the ground is stable. Otherwise, it may fail. A value of SF ≥ 1.0 is required depending on the application.

2) Estimation of rock mass constants

To use the above four equations of 5.10, rock property σ_c and rock mass constants a, s, m_b and m_i are required.

Ideally, both σ_c and m_i should be determined by tests on rock samples. Under the uniaxial test, σ_c is obtained by averaging the test results from a number of rock samples of the same type of rock. Under triaxial tests, m_i can then be determined from a number of tests at various confining pressures by linear regression analysis using the following equation, which is derived from Eqn. 5.10a

$$(\sigma_1' - \sigma_3')^2 = m_i \sigma_c \sigma_3' + \sigma_c^2 \qquad (5.13)$$

Unfortunately, it is often not possible to conduct triaxial tests in every case and for every type of rock mass. Hoek et al. (1995) compiled many case studies and came up with suggested values of σ_c and m_i. For convenience, these results are summarized in Tables 5.1 and 5.2, respectively, as references. However, it is important for users to read the original publications for detail and more accurate descriptions.

The values of a, m_b and s can be determined either using the Table in Hoek et al. (1995), or by the following three equations

$$m_b = m_i \exp((GSI\text{-}100)/28) \qquad (5.14)$$

Eqn. 5.14 and the following two equations correlate the rock mass parameter with a rock mass index (GSI). GSI is the

Geological Strength Index used in engineering practice to describe the rock mass quality. Its value ranges from approximately 10 for extremely poor rock masses to 100 for intact rock. GSI will be discussed in more detail later.

Table 5.1 Summary of estimated σ_c (After Hoek et al. 1995).

Rock quality[*]	Description in field estimate	σ_c (MPa)	$I_{s(50)}$ (MPa)
R6, Extremely strong	Repeated hammer blows can only chip rock	> 250	> 10
R5, Very strong	Requires many blows of hammer to break intact rock	100-250	4-10
R4, Strong	Specimens broken by a single hammer blow	50-99.9	2-3.9
R3, Medium strong	Firm blow with geological pick will indent rock to 5mm, knife just scrapes surface	25-49.9	1-1.9
R2, Weak	Knife cuts rock but too hard to shape into triaxial specimens	5-24.9	-
R1, Very weak	Rock crumbles under firm blows of geological pick and can be shaped with knife	1-4.9	-
R0, Extremely weak	Indented by thumbnail	0.25-0.9	-

* Note: Grade as per ISRM 1981.

Depending on the GSI value, constants a and s are determined by two separate formulae.

If GSI > 25 (undisturbed rock masses, where excavation will change very little, or not at all, the original rock mass condition),

$$\begin{cases} s = \exp(\,(GSI\text{-}100)/9) \\ a = 0.5 \end{cases} \quad (5.15)$$

If GSI < 25 (undisturbed rock masses, where excavation will alter the original condition of the rock mass, e.g., making it loose, highly fractured, or alter the confinement),

$$\begin{cases} s = 0 \\ a = 0.65 \text{ - } GSI/200 \end{cases} \qquad (5.16)$$

Now the key is to determine the GSI value for a specific application, to be discussed later.

Table 5.2 Summary of estimated m_i values for intact rocks (After Hoek et al. 1995).

Rock type	Group	Estimate of m_i
Sedimentary	Clastic	Varies with texture: from 4 (very fine) to 22 (coarse). Examples: 18 (Greywacke), 19 (Sandstone), 9 (Siltstone), 4 (Claystone)
	Nonclastic Organic	Uniform. Examples: 7 (Chalk), 8 – 12 (coal)
	Carbonate	Varies with texture. Example: 8 (limestone)
	Chemical	Varies with texture. Examples: 16 (Gypstone), 13 (Anhydrite)
Metamorphic	Non-foliated	Varies with texture: from 9 (coarse) to 24 (fine). Examples: 9 (Marble), 24 (Quartzite)
	Slightly foliated	Varies with texture: from 6 (fine) to 30 (coarse). Example: 31 (Amphibolite)
	Foliated*	Varies with texture: from 9 (very fine) to 33 (coarse). Examples: 9 (Slate), 33 (Gniess)
Igneous	Light	Varies with texture: from 16 (fine) to 33 (coarse). Examples: 19 (Andesite), 33 (Granite), 16 (Rhyolite)
	Dark	Varies with texture: from 17 (fine) to 27 (coarse). Examples: 22 (Norite), 27 (Gabbro)

* These values are for intact rocks tested normal to foliation. They may be very different in other directions.

3) Shear strength of jointed rock masses

At this point, it is necessary to introduce a set of equivalent equations for determination of cohesion, c and friction angle, ϕ to estimate shear strength in rock masses. In slope design and even in underground drifts design, the equilibrium analysis is often expressed in terms of the shear failure criterion, such as the Mohr-Coulomb criterion. In this case, the normal and shear stresses (σ_n, τ) acting on a plane are expressed in terms of the principal stresses:

$$\sigma_n = \sigma_3 + \frac{\sigma_1 - \sigma_3}{\partial\sigma + 1} \tag{5.17}$$

$$\tau = (\sigma_n - \sigma_3)\sqrt{\partial\sigma} \tag{5.18}$$

where $\partial\sigma$ is a function varying with the rock mass constants.

If GSI > 25, when a =0.5:

$$\partial\sigma = 1 + \frac{m_b\,\sigma_c}{2\,(\sigma_1 - \sigma_3)} \tag{5.19}$$

If GSI < 25, when s =0:

$$\partial\sigma = 1 + a\,m_b{}^a \left(\frac{\sigma_3}{\sigma_c}\right)^{a-1} \tag{5.20}$$

For a given type of rock mass and ground condition, the values of a, m and s can be determined as above. When a number of (σ_n, τ) values are calculated at various stress levels by Eqns. 5.17 and 5.18, they can be plotted on a σ - τ plane to determine the failure envelope, as demonstrated in Fig. 5.8.

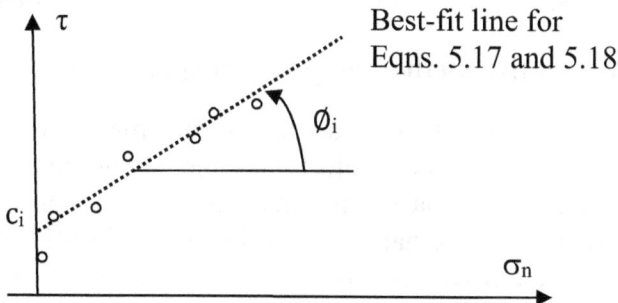

Fig. 5.8 Determination of average friction angle and cohesion.

The average values of cohesion c_i and friction angle ϕ_i can be determined by linear regression analysis through a best-fit line of the plotted (σ_n, τ) values. If the failure envelope is best represented by a curve, a tangent line through a point on the envelope at a specific stress level can be used as an approximation within a corresponding stress range.

4) Estimation of GSI values

GSI is the single most important parameter for the empirical failure criterion. As there have been numerous works completed on rock mass classification, for engineering design purpose it only makes sense to link GSI with those indices. The most common rock mass classification indices (which will be explained in detail in the next chapter) in use today is the RMR in the CSIR system and the Q in the NGI system. The most up to date and commonly used values of these two indices are used below.

Based on the RMR values updated in Bieniawski (1989),

$$GSI = RMR_{89} - 5, \text{ for } RMR_{89} > 23 \tag{5.21}$$

If $RMR_{89} < 23$, the RMR value cannot be used to determine the GSI and the modified Q value (with $J_w = 1.0$ and $SRF = 1.0$) should be used:

$$GSI = 9 \ln Q' + 44 \tag{5.22}$$

$$\text{where } Q' = \frac{RQD}{J_n} \frac{J_r}{J_a} \tag{5.23}$$

RQD is another rock quality index. J_n, J_r, J_a, J_w and SFR are part of the Q system, all of which will be discussed in next chapter.

5) An alternative method for estimating m and s

It happens often that there is no data of rock mass classification available in a specific project and the above methods for estimating the rock mass constants cannot be implemented. In the following, an alternative method suggested by Hoek et al. (1995) to determine the rock mass constants is presented. Selection of m, s values is described in detail in a comprehensive table. The rock mass is classified based on its structures and the surface condition of the discontinuities within the rock mass. The definition of rock mass category and quality of joint surface conditions are summarized in Table 5.3. Users are however encouraged to read the original publications for detail and descriptions.

Based on these defined terms, for users' convenience the suggested values of m_b/m_i ratio and s, plus GSI are presented here in a graphical format in Figs. 5.9 to 5.11, respectively. To use this method, users first should have adequate geotechnical information in the field about the rock mass and discontinuities. Based on your own judgement, you should choose the proper category for the specific rock mass and discontinuity and then select a proper value for m_b/m_i and s from Figs. 5.9 to 5.10.

If no rock mass classification is performed for a specific site, Fig. 5.11 can be used to estimate GSI based on the rock structures and joint surface conditions. GSI values range from 10 for the Crushed rock with "Very poor" joint surface condition to 85 for the Blocky rock with "Very good" joint surface condition.

Table 5.3 Category of rock masses and quality of joint surface conditions for use of selecting m and s constants (After Hoek et al. 1995).

a) Category of rock mass based on structures

Class	Descriptions
Blocky	Very well interlocked, undisturbed rock mass consisting of cubic blocks formed by three orthogonal discontinuity sets.
Very blocky	Interlocked, partially disturbed rock mass with multi-faceted angular blocks formed by four or more discontinuity sets.
Blocky/seamy	Folded and faulted with many intersecting discontinuities, forming angular blocks.
Crushed	Poorly interlocked, heavily broken rock mass with a mixture of angular and rounded blocks.

Table 5.3 (continued)
b) Quality of joint surface conditions

Quality	Descriptions
Very good	Very rough, unweathered surfaces
Good	Rough, slightly weathered, iron stained surfaces.
Fair	Smooth, moderately weathered or altered surfaces.
Poor	Slickensided, highly weathered surfaces with compact coatings or fillings containing angular rock fragments.
Very poor	Slickensided, highly weathered surfaces with soft clay coatings or fillings.

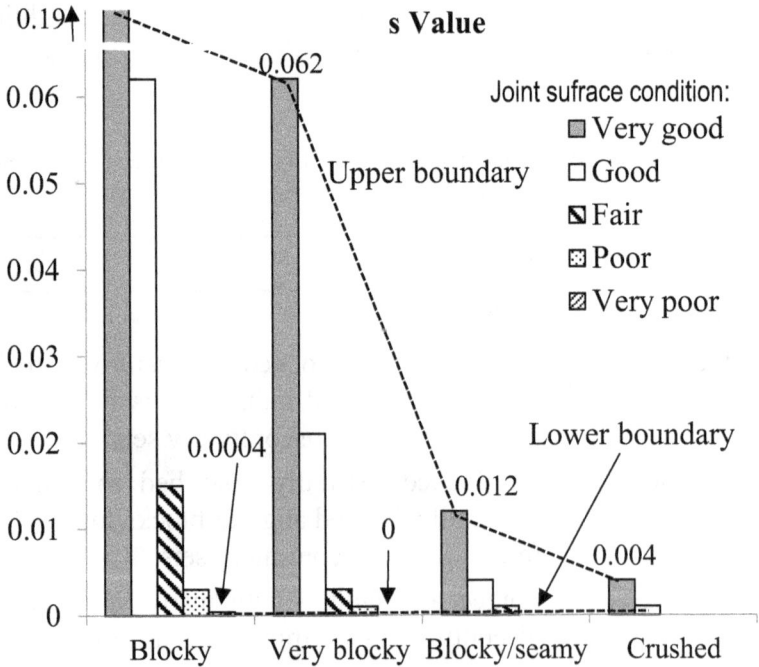

Fig. 5.9 S value for jointed rock masses (After Hoek et al. 1995).

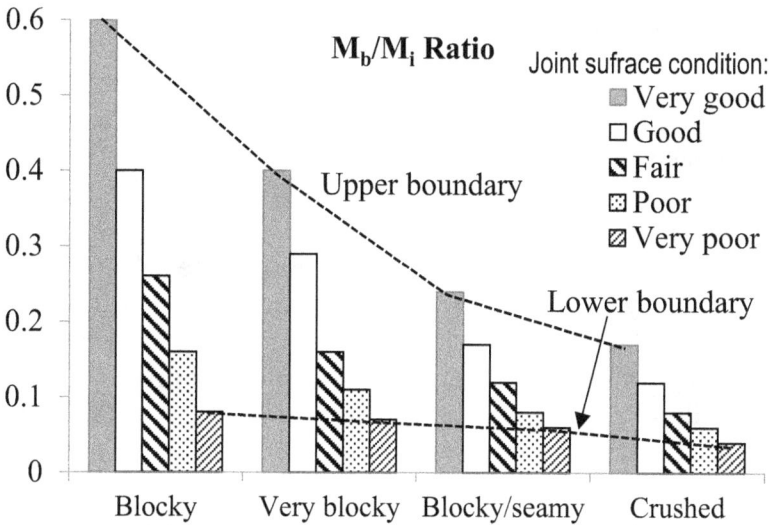

Fig. 5.10 M_b/M_i ratio for jointed rock masses (After Hoek et al. 1995).

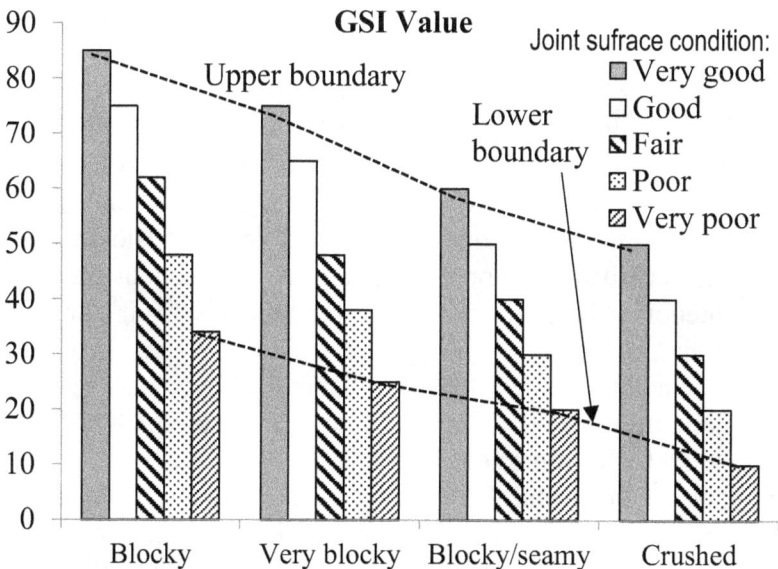

Fig. 5.11 GSI value for jointed rock masses (After Hoek et al. 1995).

Notes and Discussion

1. A comparison of the data presented in Figs. 5.9 to 5.11 and those in the earlier version of the empirical failure criterion (Hoek and Brown 1980a) reveals that the method of selecting values for m and s is quite different after having been refined over many years.
2. For the three parameters, m_b/m_i ratio, s and GSI, an upper boundary and a lower boundary are established for the included rock mass conditions. These two boundaries define a range of values to use. This would allow for proper selection of these parameters for a specific site, which may not fit exactly the rock categories and joint surface conditions as described in Table 5.3.
3. The m_b/m_i ratio in Fig. 5.10 ranges from a minimum of 0.04 to a maximum of 0.6. The effect of rock mass structures is reflected in the rock mass categories from Blocky to Crushed. The effect of joint surface conditions is reflected within each category. In application, if a specific site condition falls between two categories and the joint surface conditions fall between two defined conditions, a proper value may be selected by interpolation in the chart.
4. The s values in Fig. 5.9 range from a minimum of 0 to a maximum of 0.19. This value is far below 1.0 (for intact rock). Even with "Very good" and "Good" joint surface conditions, the s value is extremely low for Blocky/seamy and Crushed categories and it is set to zero in these two categories if the joint surface condition is "Fair" or below "Fair".
5. As an alternative to Items 3 and 4 above, one may first estimate the GSI value and then use Eqns. 5.14 to 5.16 to estimate the values of m_b and s.
6. As shown in the three Figs. 5.9 to 5.11, the rock category only covers jointed rock masses with minimum three sets of orthogonal discontinuities. The cases of two sets of joints or three sets of non-orthogonal joints are not specified. In that case, the m and s values may be interpreted between the intact rock and the Blocky rock, provided the joints do not play a dominant role in the behaviour of the rock mass.

In this case, one suggestion to approach the problem is to treat it in two different ways: a) consider the rock mass as jointed rock mass as above, and b) consider the joint sets separately to come up with another solution, then compare the outcomes from both approaches and finally make a decision based on the "worst" scenario. This is only a suggestion by the author of this book as a starting point in the design process. Users must be cautious in doing so with consideration to other relevant factors as it is yet to be verified in practice.

General limitations of the empirical failure criterion

It is important to know where the empirical failure criterion can be used. In general,

- it is only applicable to isotropic intact rock or highly jointed rock mass and rock mass with many random joint sets, which can be considered as homogeneous and isotropic but at a much reduced quality level.
- it cannot be applied to a rock mass which contains only one joint set, or is controlled by a single discontinuity set, such as bedding planes. These discontinuities must be treated individually. When there are two sets of joints, the criterion may be used with extreme care, provided neither of the joints have soft coating nor dominate the behaviour of the rock mass.
- the effective normal stress in the field should be $\sigma' \le \sigma_c$
- the size of an opening and the joint spacing are not considered in this process. Use with care.

Example 5.3

A rock mass consists of primarily Rhyolite subjects, slightly weathered, with several joint sets at an average 2 m spacing. Uniaxial compressive strength is $\sigma_c = 120$ MPa. In-situ stresses in the field are estimated as $\sigma_1 = 80$ MPa, $\sigma_3 = 20$ MPa. Underground excavation will slightly disturb the rock.

a) If the confining pressure $\sigma_3 = 20$ MPa, what is the failure pressure of a joint free specimen in triaxial test according to the empirical criterion?

b) If rock mass classification gives a RMR_{89} value of 65, what is the rock mass strength by the empirical failure criterion?

c) If there is no rock mass classification data available, assess the ground stability after excavation of an underground chamber. The maximum stress around the chamber after excavation is estimated to be twice the maximum in-situ stress.

Solutions:

a) σ_c is already given. We need to determine m_i to use this criterion. For intact rock of Rhyolite, from Table 5.2, the estimated m_i value is 16.

For a joint free intact rock specimen, $s = 1$ and Eqn. 5.10a is used with no water pressure

$$\sigma_1' = \sigma_3' + \sqrt{m_i \sigma_c \sigma_3' + \sigma_c^2}$$

$$= 20 + \sqrt{16 \times 120 \times 20 + 120^2}$$

$$= 225 \text{ (MPa)}.$$

The result indicates that intact rock strength in triaxial compression is greater than the uniaxial strength σ_c.

b) In this question, rock mass classification index is provided. We can use the available equations to determine the required parameters.

By Eqn. 5.21,

$$GSI = RMR_{89} - 5 = 65 - 5 = 60.$$

By Eqn. 5.14,

$$m_b = m_i \exp((GSI-100)/28)$$

$$= 16 \exp((60-100)/28)$$

$$= 3.83.$$

By Eqn. 5.15 (because GSI > 25), a = 0.5 and

$$s = \exp(\frac{GSI\text{-}100}{9}) = \exp(\frac{60\text{-}100}{9})$$

$$= 0.01174.$$

Use Eqn. 5.10b without water pressure and consider the minor principal stress in the field σ_3 = 20 MPa,

$$\sigma_1' = \sigma_3' + \sqrt{m_b\,\sigma_c\,\sigma_3' + s\,\sigma_c^2}$$

$$= 20 + \sqrt{3.83 \times 120 \times 20 + 0.01174 \times 120^2}$$

$$= 116.7\ (\text{MPa}).$$

This is the rock mass strength under the specified field stresses.

c) In this question, we are required to assess the ground stability after excavation. It means we need to determine the strength of the rock mass and compare it with the maximum stress after excavation.

The maximum stress after excavation, as specified in the question, is

$$2\sigma_1 = 2\times80 = 160\ (\text{MPa})$$

To estimate the rock mass strength with the empirical failure criterion, values of m_b, a and s must be estimated for the rock mass based on the description. When there is no GSI or other rock mass classification information available, they cannot be determined using Eqns. 5.13 - 5.16. We will use an alternative method with information available in Figs. 5.9 to 5.11.

First, compare the field condition with those defined in Table 5.3. In rock mass "structure", "several joint sets with 2 m joint spacing and slight disturbance from excavation seem to match the category of "Very blocky – partially disturbed".

In "joint surface condition", "slightly weathered" in hard rock Rhyolite, which would have rough surface, seems to match the "Good" quality.

Therefore, for the rock mass in the field we have GSI = 65 from Fig. 5.11.

Then by Eqn. 5.15, a = 0.5 and

$$s = \exp(\frac{GSI\text{-}100}{9}) = 0.02.$$

(s can also be estimated from Fig. 5.9, approximately 0.02).

By Eqn. 5.14,

$$m_b/m_i = \exp(\frac{GSI\text{-}100}{28}) = 0.287.$$

(m_b/m_i ratio can also be estimated from Fig. 5.10, approximately 0.29). Because m_i was estimated earlier to be 16,

∴ $m_b = 0.29\ m_i = 0.29 \times 16 = 4.64.$

By Eqn. 5.10b for jointed rock mass,

$$\sigma_1' = 20 + \sqrt{4.64 \times 120 \times 20 + 0.02 \times 120^2}$$
$$= 127\ (MPa).$$

This indicates that the strength of the rock mass under a triaxial compressive stress field is approximately the same as σ_c.

The ratio of the rock mass strength over the field major principal stress after excavation is:

$\sigma_1' / (2\sigma_1) = 127 / 160 = 0.79$, which is < 1.0.

The rock mass is therefore unstable after excavation because the stress concentration is very high and most likely a fracture zone will develop around the opening.

6) Strength of schistose and layered rocks

Schistose and layered rocks include those rocks which have one set of joints or very weak planes, such as slates, shales, etc. They are naturally transversely isotropic. For this type of rock, the failure greatly depends on the stress condition in the field and on the loading condition in laboratory tests. It may take place through the rock mass or a weak discontinuity.

As shown by Eqn. 5.7b and Fig. 5.4, the most vulnerable plane is that which has an angle $\Psi = 45° - \phi_i/2$ with respect to the major stress direction. If a weak plane happens to be near that direction, it will dominate the rock strength and shear failure of the rock will be on that weak plane. As a result, failure of the rock will take place at a lower stress level, as illustrated in Fig. 5.12. A weak plane's position Ψ with respect to the major stress direction may therefore have significant effect on the rock strength. This effect exists in an angular range with $\Psi=10°\sim50°$ approximately from the major stress direction, forming a trough in Fig. 5.12. The width of the trough depends on the difference between the internal rock shear strength (c_i, ϕ_i) and the shear strength on the joint plane (c, ϕ). The larger the difference, the wider the trough.

If triaxial tests are conducted at various confining pressures, there will be a series of similar curves, shifting higher as σ'_3 increases (Fig. 5.12).

If there are several sets of joints, each with different characteristics, there will be several similar curves, shifting slightly left or right as the joint angle Ψ changes. These curves will overlap each other. The joint with the lowest shear strength will dominate the rock mass failure.

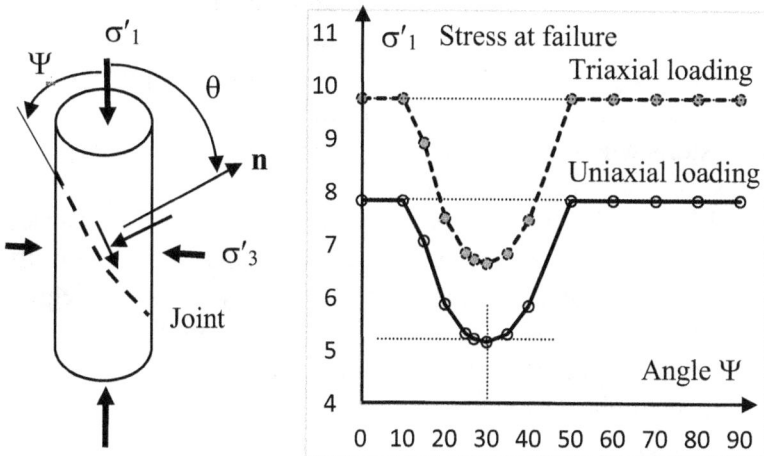

Fig. 5.12 The effect of a weak joint position (Ψ) on the failure stress of a rock specimen in compression test.

Chapter 6

Rock Mass Classifications

It needs to be pointed out that classification of rock masses is not simply for the purpose of classification, such as, "black rock" or "pink rock", rather it is for the purpose of engineering design. Rock mass classification is intended to provide engineers with a quantitative tool to measure the complex rock mass system in the field.

An underground excavation is a complex engineering structure. In many cases the only theoretical tools available to an engineer to help complete this task are a number of grossly simplified models of certain processes which interact to control the stability of the excavation. These models can generally only be used to analyse the influence of one individual process at a time, such as the influence of geological discontinuities or of high rock stress upon the excavation. It is seldom possible theoretically to determine the interaction of these processes. The designer faces the challenge of making a number of design decisions where engineering judgement and practical experience must play an important role.

If an engineer has experience in design and/or construction of underground excavation in a certain type of rock condition, his / her design decisions may be made with some degree of confidence with similar rock conditions. On the other hand, where no such experience is readily available, what criteria can be used to check whether one's decisions are reasonable? How can one judge whether the span of a drift is too large or whether too many or too few rock bolts have been specified in that ground condition?

The answer lies in some sort of common classification system for rock masses, which can enable an individual to relate one's own set of site conditions to the conditions encountered by others. Such a classification system acts as a vehicle which enables a designer to relate the experience on rock conditions and support requirements gained on other sites to the conditions anticipated on his/her own site. Development of various classification systems for underground support is well documented in literature.

In underground excavation, there are numerous factors which will influence the ground stability. These include geological structures, rock mass properties, field stresses, water conditions, excavation and support methods, size and geometry of an opening, etc. They all exist simultaneously and interact with one another in controlling ground stability. Some may play a more important role than others. They all have to be considered together in order to make a safe and reasonable design.

An engineering design needs to be judged for its safety and cost effectiveness, in addition to its functionality. The judgement will be based on:
- the factors mentioned above,
- the previous experience in similar conditions,
- comparison between what has been done and what is to be done.

In the early days of developing rock mass classification systems, there was only qualitative description due to the complexity of the rock mass itself. Today it has been developed into quantified systems.

A number of rock mass classification systems have been developed in the past. Each one of them was developed primarily for a specific type of applications and therefore more emphasis was put on particular parameters. As a result, each has some limitations in its application. In rock engineering, several classification systems are commonly used today and they will be discussed in the following sections.

6.1 Terzaghi's Classification System

This classification system was initially developed by Terzaghi (1946) based on his experience in steel-supported railroad tunnels. It is a simple system primarily for estimating the load to be supported by steel arches in tunnels. This is a very important step. In this system, various types of ground conditions are described and a range of rock loads are assigned for each ground condition. The intention was to quantify the author's experience in such a way that it could be used by others. It has been widely used in tunnelling around the world since it was published.

Terzaghi stressed the importance of geotechnical investigation which should be carried out before a tunnel design is completed, particularly the importance of obtaining information on defects in the rock mass. He stated that *"From an engineering point of view, a knowledge of the type and intensity of the rock defects may be much more important than the type of rock which will be encountered"*.

Under this system, the rock mass is divided into several categories, as follows:

1. Intact rock: Rock mass in this category contains no joints or hair cracks. When it breaks, it breaks across sound rock. On account of the injury to the rock due to blasting, spalls may drop off the roof several hours or days after blasting, a condition called *spoiling*. Hard, intact rock may also involve spontaneous and violent detachment of rock slabs from the sides or roof, a condition called *popping*.

2. Stratified rock: This type of rock mass consists of individual strata with little or no resistance against separation along the boundaries between strata. The strata may or may not be weakened by transverse joints. In such rock, spalling conditions are quite common.

3. Moderately jointed rock: Rock mass contains joints and hair cracks, but the blocks between joints are locally bonded together or so intimately interlocked that vertical walls do not require lateral support. In this type of rock, both spalling and popping conditions may be encountered.

4. Blocky and seamy rock: Rock in this category consists of chemically intact or almost intact rock fragments which are entirely separated from each other and imperfectly interlocked, In such rock, vertical walls may require lateral support.

5. Crushed but chemically intact rocks: These rocks have the characteristics of crusher run. If most or all of the fragments are as small as fine sand grains and no re-cementation has taken place, crushed rock below the water table exhibits the properties of a water-bearing sand.

6. Squeezing rock: This type of rock, if a tunnel is in place, will slowly advance into the tunnel without a perceptible volume increase. A prerequisite for squeeze is a high percentage of microscopic and sub-microscopic particles of micaceous minerals or of clay minerals with a low swelling capacity.

7. Swelling rock: This type of rock will advance into the tunnel mostly by expansion. The capacity to swell seems to be limited to those rocks that contain clay minerals such as montmorillonite, with a high swelling capacity.

The diagram in Fig. 6.1 illustrates the concept of the classification system. The major factors considered in the system include:

- mechanical defects of the rock mass,
- dead load above the support,
- water sensitivity of the rock mass.

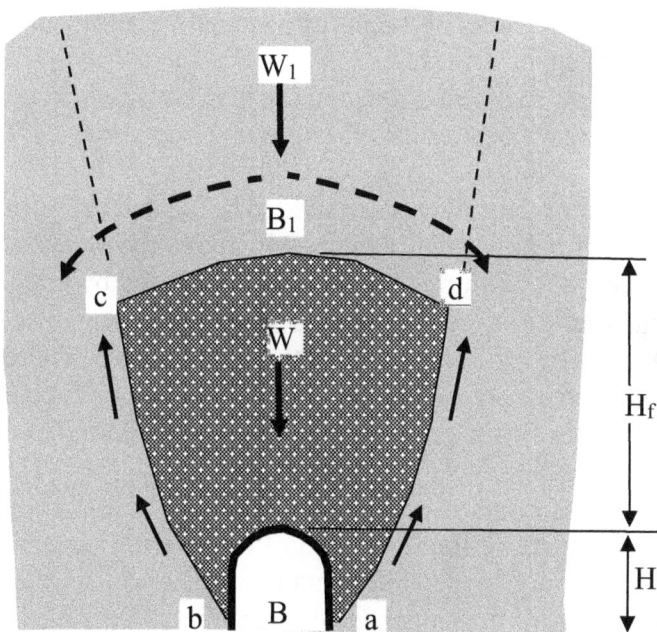

Fig. 6.1 Simplified diagram for loosened rock above a tunnel (After Terzaghi 1946).

It is also assumed in this classification system that:
- excavation of the tunnel causes stress relaxation and loosening of rock in the ground,
- the fractured rock is within the area enclosed by a-b-c-d (in Fig. 6.1) between the loosened rock and the intact rock.
- the weight of the overburden (above H_f elevation and between the two dashed lines), W_1, is transferred to the sides through a stress arch effect, as demonstrated by the dashed arrows.
- the tunnel only needs to support the unbalanced weight of the loosened dead rock W, which has height H_f.

Therefore, tunnel support design is focused on estimation of height H_f of the fractured rock. H_f is defined as a function of a number of factors,

$$H_f = f(B_1, H, B \text{ and rock mass properties})$$

Based on numerous tests on steel supports, a classification system was proposed and an estimation of H_f is provided for each type of ground condition corresponding to the categories described earlier, as shown in Table 6.1. In each condition, the estimated rock load on tunnel support is expressed in terms of the tunnel span and height.

The first four conditions are for mechanical defects and the last two for water sensitivity.

A few points should be noted when using Table 6.1:
- the roof of the tunnel is assumed to be located below the water table. If it is located permanently above the water table, the values for conditions 6 and 7 can be reduced by up to 50%.
- real shale in an un-weathered state is no worse than other stratified rocks. However, if the term shale refers to firmly compacted clay sediments which have not yet acquired the properties of rock, such shale may behave like squeezing or even swelling rock.
- If a rock formation consists of a sequence of horizontal layers of sandstone or limestone and of immature shale, excavation will commonly encounter gradual compression of the rock on both sides of the tunnel and downward movement of the roof.

Table 6.1 Summary of Terzaghi's rock load on tunnels (After Terzaghi 1946).

Rock Mass Condition	Height of Loosened Rock, H_f [feet]	Notes
1. Hard and intact	0	Light lining required only if spalling or popping occurs.
2. hard stratified, or schistose	0 to 0.5 B	Light support, mainly for protection against spalls.
3. Massive, or moderately jointed	0 to 0.25 B	Load may change erratically from point to point.
4. Blocky and seamy	0.25B~1.10 (B+H)	Little or no side pressure.
5. Crushed but chemically intact	1.10 (B + H)	Considerable side pressure.
6. Squeezing rock	(1.1~4.5) (B+ H), increasing with depth	Heavy side pressure, invert struts required. Circular ribs are recommended.
7. Swelling rock	Up to 250', irrespective of (B+ H)	Circular ribs are required. In extreme cases use yielding support.

Applications

For excavations in shallow depth, this system provides a good design tool. However, since it was based on steel arch support and not very deep conditions, a few points should be kept in mind when used for mining related applications:

- Because the original tests were based on steel arch-supported tunnels, the results may not be appropriate for other types of support, such as shotcrete and rock bolts.
- Due to lack of quantitative information on rock properties and field conditions, there is some difficulty in making a judgement without experience.

- There are differences between the load on a shallow tunnel and that on an underground drift. Stress is much higher underground than near the surface.

Lauffer (1958) extended the system to estimate a free stand-up time without any support for different types of ground conditions and different active tunnel spans. Since support is usually required in most applications in the ground conditions encountered in practice, the free stand-up time will give some idea of when and how fast ground support is to be provided after excavation. This is important for excavations in soft and weak ground. For hard rock, the free stand-up time may help decide whether ground support is required or what type of support is to be used.

6.2 Rock Quality Designation (RQD)

This system is solely based on the rock core recovery from diamond drilling. It gives a quantitative description and makes it easier to implement. It is usually part of core logging and has therefore been widely used in rock engineering. It can be used alone as an indicator of the integrity of rock mass and is also a major component in other rock mass classification systems.

This Rock Quality Designation (RQD) was proposed by Deere (1964) and is widely used even today. The RQD value is defined as the percentage of core recovered in intact pieces of 100 mm (4") or longer in length, over the total length of the logged borehole, i.e.,

$$RQD = \frac{\text{total length of core pieces with length} \geq 100 \text{ mm (4")}}{\text{total length of borehole in which core is recovered}}(\%) \quad (6.1)$$

It is normally accepted that RQD should be determined on a core of at least 50 mm diameter, which should have been drilled with double barrel diamond drilling equipment. A RQD value would usually be established for each core run of 2 m or so. This determination is simple and quick, and it is now often carried out in conjunction with the normal geological logging of cores.

Depending on the RQD value, rock masses are classified into five classes, as shown in Table 6.2.

Table 6.2 Rock mass classification based on RQD values.

RQD (%)	Rock Quality	length > 4"
		10
< 25	Very poor	4.5
25 - 50	Poor	
50 - 75	Fair	8.6
75 - 90	Good	
90 - 100	Very good	12.4
		6
		6.5

Example 6.1

For the core recovery shown in Table 6.2, the total core recovered is 55" and the total length of hole drilled and logged is 60". Determine the RQD value of the hole.

Solutions:

From Table 6.2, the total length of the core pieces ≥ 4" is 48". Since the whole length of the drilled hole is logged, the denominator is 60". By Eqn. 6.1,

RQD = 48/60 = 80%

The rock mass has "Good" quality.
    There are a few points to be noted about the RQD system:
• It does consider joint spacing, but not joint orientation and continuity, which are very important.

- It does not account for the thin weak infillings (such as clay) in joints, which can be encountered both near the surface and at depth and reduce the frictional resistance along the joints significantly.
- It does not provide an adequate indication of the range of behaviour patterns, which may be encountered when excavating underground.

Application of RQD in tunnel support estimate

Based on tunnel width and RQD value, it is possible to estimate the support requirements for the tunnel. Merritt (1972) proposed to use RQD to evaluate the support requirement of tunnels with rock bolts and steel rib. Some examples are given in Table 6.3.

Table 6.3 Examples of use of RQD for tunnel support (After Merritt 1972).

RQD(%)	Span: ≤3m	3~7m	7~10m	10~15m
≤10	Bolting* or Steel rib	Steel rib ⟶		
10~25	Bolting	Bolting, or Steel rib	Steel rib ⟶	
25~50	Light support or Bolting	Bolting	Bolting or Steel rib	Steel rib
50-75	None or light support	Light support or Bolting	Bolting	Bolting or Steel rib
75~100	None or light support	⟶	Light support or Bolting	Bolting

* Suggested bolting is 4' to 6' centered pattern.

The 4' × 4' or 6' × 6' bolting pattern is widely used today in industry. However, the data shown in Table 6.3 has serious limitations in areas where joints have thin clay fillings. Therefore, great caution should be exercised in use of the suggested support.

6.3 Geomechanics Classification System for Jointed Rock Mass (RMR)

This classification system was once known as the CSIR (South Africa Council for Scientific & Industrial Research) system. Lately it is simply called the Rock Mass Rating (RMR) system. It was initially proposed by Bieniawski (1976) and updated a few years later (Bieniawski 1989). Over the years, this system has been successively refined as more case records became available. Over the course, Bieniawski has made *significant changes* in the ratings assigned to individual parameters. Discussions and data used in this textbook are based on the 1989 version. Unless specified otherwise, the values will be designated as RMR_{89}. Values from other versions may differ.

It has become clear to us from previous discussions that:
- the behaviour of rock mass underground is complex,
- there are numerous factors which influence the rock mass,
- all factors act as an integrated system and no single simple index alone is adequate to describe the rock mass.

Furthermore, a simple useful classification system should combine the major factors together to reflect the real rock mass. Influence of factors, such as clay filling and weathering should also be included.

To be practical, a classification system should also:
- group the rock mass by their behavior,
- help understand rock mass characteristics,
- use quantitative data in the process,
- facilitate planning and design in practical use,
- be simple to use and be able to be implemented by users who have less real-life experience.

This RMR system is intended to satisfy those requirements. In this system, major factors are grouped together and a score is assigned to each group based on certain criteria. Then the total rating of the rock mass is summed up to determine the overall quality of the rock mass. The system can be given as follows:

$$RMR = \sum_{i=1}^{5} R_i + R_j \qquad (6.2)$$

where, R_i is the rating value of parameters in group i (to be described below), $R_i \in [0, 30]$, and R_j is the adjustment rating for joint orientation, $R_j \in [-60, 0]$.

Description of individual parameters

The following is based on the work of Bieniawski (1989).

1) R_1, rock mass rating by intact rock strength

The strength of intact rock can be determined by uniaxial compression test on rock specimens in laboratory or by point load test in laboratory or in the field. R_1 values are assigned based on rock strength, as shown in Table 6.4.

Table 6.4 Rock mass rating by intact rock strength.

σ_c (MPa)	>200	200 - 100	100 - 50	50 - 25	25 - 10	10 –3	3 - 1
$I_{s(50)}$ (MPa)	>8	4-8	2-4	1-2	-	-	-
R_1*	15	12	7	4	2	1	0

* Note: If a value falls at the boundary between two choices, an average of the two may be used, although this is not expected to have a significant effect on the end result of classification.

When $I_{s(50)} < 1$ MPa, use of I_s is not recommended and σ_c should be used.

2) R_2, rock mass rating by RQD value

RQD is determined from core logging as discussed earlier and a rating is given in a range of 3 to 20 (Table 6.5).

It needs to point out that RQD may be determined for the whole borehole or just for individual sections of the borehole. In the latter, a section usually responds to a type of rock mass or a domain where rock mass quality is to be assessed.

Table 6.5 Rock mass rating by RQD value.

RQD (%)	100 - 90	90 - 75	75 - 50	50 - 25	< 25
R_2	20	17	13	8	3

3) R_3, rock mass rating by joint spacing

The average spacing between two adjacent joints in a set of joints is estimated based on measurements in the field. Joints may include any bedding plane, fault or other discontinuities. Ratings are specified in Table 6.6.

Table 6.6 Rock mass rating by joint spacing.

Joint spacing (m)	> 3	3 - 1	1 - 0.3	0.3 - 0.05	< 0.05
R_3	20	15	10	8	5

4) R_4, rock mass rating by joint surface condition

This factor considers joint length (continuity), separation or aperture, surface roughness, infilling materials and weathering condition. The suggested ratings are assigned based on descriptions of the surface conditions, as shown in Table 6.7a.

Table 6.7a Rock mass rating by joint surface condition.

Joint surface condition	Very rough, Not continuous, No separation, Unweathered wall rock	Slightly rough, Separation <1mm, Slightly weathered walls	Slightly rough, Separation <1mm, Highly weathered walls	Slickensided or Gouge <5mm thick, or Separation 1-5mm, Continuous	Soft gouge >5mm thick, or Separation >5mm, Continuous
R_4	30	25	20	10	0

If quantitative information on joint surface condition is available, R_4 rating may also be calculated by summing up the ratings of individual parameters, which are given in Table 6.7b.

Table 6.7b Detail for rock mass rating by joint surface condition*.

a. **Discontinuity** length (m)	< 1	1 - 3	3 - 10	10 - 20	> 20
Rating	6	4	2	1	0

b. **Separation** (mm)	none	< 0.1	0.1 - 1.0	1 - 5	> 5
Rating	6	5	4	1	0

c. **Roughness**	Very rough	Rough rough	Slightly	Smooth	Slickensided
Rating	6	5	3	1	0

d. **Infilling** (gouge)	None	Hard < 5mm	Hard > 5mm	Soft < 5mm	Soft > 5mm
Rating	6	4	2	2	0

e. **Weathering condition**	Un-weathered	Slightly weathered	Moderately weathered	Highly weathered	Decomposed
Rating	6	5	3	1	0

* Add up all five ratings to determine R_4.

5) R_5, rock mass rating by water condition

Presence of water or water pressure could drastically alter the behaviour and the properties of a rock mass. The ground water condition may be represented by water inflow rate, water pressure or field observation. Rating are assigned in Table 6.8.

Table 6.8 Rock mass rating by water condition*.

Inflow rate (L/min./10m length of drift)	None	< 10	10 - 25	25 - 125	> 125
Ratio of water pressure/σ_1	0	< 0.1	0.1 - 0.2	0.2 - 0.5	> 0.5
Observation	Dry	Damp	Wet	Dripping	Flowing
R_5	15	10	7	4	0

(* If there is more than one choice, choose the worst condition.)

6) R_j, rock mass rating adjustment by joint orientation

The drift orientation with respect to that of a joint is very important. A joint may have very little or serious effect on ground stability depending on its relative orientation. Rating values are assigned following a two-step procedure:
Step 1. Referring to joint strike and drift direction as shown in Fig. 6.2, assess the influence of joint orientation relative to the drift using Table 6.9a: from very unfavourable to very favorable.

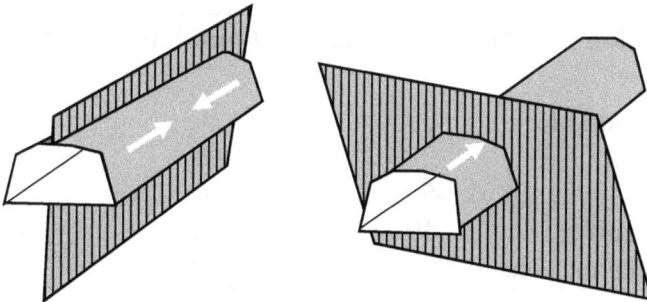

a) Joint parallel to drift   b) Joint perpendicular to drift

Fig. 6.2 Joint orientation relative to drift.

Table 6.9a Influence of joint orientation.

Joint relative to drift	Orientation Influence*
a). joint dip 0 ~ 20°	Fair
b). joint dip > 20° and joint strike // drift axis:	
joint dips 20° – 45°	Fair
joint dips 45° – 90°	Very unfavourable
c). joint dip > 20° and joint strike ⊥ drift axis:	
joint dips 20° – 45°:	
drift drive with dip	Favourable
drift drive against dip	Unfavourable
joint dips 45° – 90°:	
drift drive with dip	Very favourable
drift drive against dip	Fair

* If the joint position is between two categories, chose the "worst".

Step 2. Based on the assessed influence in Table 6.9a, determine the adjustment value on the rating using Table 6.9b. In Table 6.9b, three different applications are included: underground drifts/tunnels, foundation and slopes. Since the confinement in the field for the three applications is sequentially lessened, the negative impact of joint orientation becomes more severe and more adjustment on the rating is suggested.

Table 6.9b Rating adjustment (R_j) based on joint orientation influence.

Orientation Influence	Mine drifts/tunnels	Foundations	Slopes
Very favourable	0	0	0
Favourable	-2	-2	-5
Fair	-5	-7	-25
Unfavourable	-10	-15	-50
Very unfavourable	-12	-25	-60

Based on the total RMR value calculated by Eqn. 6.2, a rock mass is classified in one of the five classes (Table 6.10). For each class of rock mass, the unsupported stand-up time and equivalent rock properties (c, ϕ) are suggested.

Table 6.10 Rock mass classification by RMR value.

RMR value	100 - 81	80 - 61	60 – 41	40 – 21	< 21
Class	I	II	III	VI	V
	Very good	Good	Fair	Poor	Very poor
Stand-up time	20 years	1 year	1 week	10 hours	30 min.
at span (m):	15	10	5	2.5	1
Estimate of rock mass properties:					
Cohesion, c (kPa)	> 400	400-300	300-200	200-100	< 100
Friction angle, $\phi(°)$	> 45	45 - 35	35 - 25	25 - 15	< 15

Applications

Once the class of rock mass is determined, its quality index and support requirements can be estimated for use in engineering design. RMR ratings may also be used for:

a) estimation of the free stand-up time of a tunnel (see Table 6.10),
b) estimation of the equivalent cohesion and friction angle (or Mohr failure envelope) for the rock mass (see Table 6.10),
c) selection of ground support systems (more to be discussed later),
d) determination of the m and s constants for the Hoek and Brown empirical failure criteria, as discussed in the previous chapter.

Notes

1). In use of Tables 6.4 to 6.10, it is sometimes inevitable to have a parameter fall between two categories. An average of the adjacent two values may be used. For example, if σ_c is 50 MPa, a value of $(7+4)/2=5.5$ may be used for R_1. However, from experience it is not necessary to interpret the value between the lines as the end result may fall into the same category.

2). When there are multiple weak joints, their effect on the tunnel stability may be different. All effects of each joint should be assessed separately. The joint which has the lowest RMR value has the worst effect. The worst effect should be considered in design. One should never attempt to take an average of R_j values of all joints since each discontinuity has its own influence regardless of the others.

3). In field applications, the rock mass should be divided into a number of regions, with each region classified separately. The division is normally based on the geological structures, rock types, or applications.

Example 6.2

A tunnel is to be driven through slightly weathered granite. Other field information is as follows:
- a dominant joint set dip 60° with the tunnel drive direction as shown on the right,
- the joints are slightly rough and slightly weathered, with separation < 1 mm and an average 300 mm spacing,
- core logging shows RQD of 70%,
- point load test shows a strength of 8 MPa,
- the site is wet by observation.

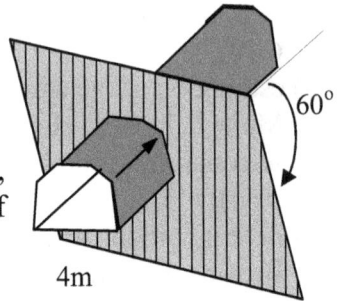

60°

4m

a) What is the RMR rating?
b) What are the estimated c and ϕ values of the rock mass?
c) What are the estimated m and s constants?
d) What is the RMR rating when the site is dry?
e) What is the RMR rating if the tunnel is // (parallel) to the joint plane, assuming the other conditions remain the same as in the description?

Solutions:

a) The selected ratings and other information are shown below:

Item	Value	Table #		Rating
Rock strength	8 MPa	6.4	R_1	12
RQD	70%	6.5	R_2	13
Joint spacing	0.3m	6.6	R_3	10
Joint condition	*	6.7	R_4	22(or 25)
Ground water	wet	6.8	R_5	7
Joint orientation	Very favourable**	6.9	R_j	0
=> total				64 (67)

* Table 6.7a gives 25 based on the description. However, if detailed information of a joint is available, Table 6.7b can be used to sum up the five individual ratings. In this case, 4 for joint

length (interpreted 1-3m), 4 for separation (<1 mm), 3 for slightly rough, 6 for no infillings, 5 for slightly weathered and the sum is 22. Either value (22 or 25) may be used.

** Table 6.9a gives "Very favourable" for driving with dip of 60°.

The total RMR = 64 and the rock mass is classified as "II, Good rock" (see Table 6.10).

b). The estimated c and ϕ values from Table 6.10 are:
$$c = 0.3 - 0.4 \text{ MPa},$$
$$\phi = 35° - 45°.$$

c). The rock mass constants for the described conditions are determined as follows:

By Eqn. 5.21, GSI = RMR_{89} – 5 = 64 – 5 = 59

By Eqn. 5.14,

$$m_b = \exp\left(\frac{GSI - 100}{28}\right) m_i = \exp\left(\frac{59 - 100}{28}\right) m_i = 0.23 \, m_i$$

where m_i is to be determined as described in the previous chapter. From Table 5.2, $m_i = 33$,

$$\therefore \quad m_b = 0.23 \times 33 = 7.59.$$

By Eqn. 5.15,

$$s = \exp\left(\frac{GSI - 100}{9}\right) = \exp\left(\frac{59 - 100}{9}\right) = 0.010.$$

d). When the site is dry with no water, $R_5 = 15$ from Table 6.8 and the others remain the same.
$$RMR = 64 - 7 + 15 = 72,$$
"II, good rock", unchanged in classification.

e). If the tunnel is driven // to the joint plane, the condition becomes "very unfavourable" (Table 6.9a), $R_j = -12$ (Table 6.9b) and the others remain the same.
$$RMR = 64 - 12 = 52, \text{ "III, fair rock", downgraded.}$$

Note: One may attempt to use the "alternative method for estimating m and s" as described in the previous chapter. It may sound logical to do so. However, that method is for jointed rock masses with a minimum of three sets of joints, while the example here has only one set of dominant joints.

6.4 NGI Rock Tunnelling Quality Index - Q

The Q index was initially proposed by Barton et al. (1974) of the Norwegian Geotechnical Institute (NGI) to determine the rock mass characteristics encountered in tunnelling and the support requirements. Like the RMR system, it has been updated many times since then (Singh et al. 1992, Grimstad 1993, Loset 1997, Grimstad et al. 2002). NGI (2013) has also published a handbook on the use of the Q-system.

The Q-system is based on a large number of case histories of underground excavations. It considers six major field factors, as shown below, and assigns a numerical rating to each factor based on measurements and description. The Q value is calculated by

$$Q = (\frac{RQD}{J_n})(\frac{J_r}{J_a})(\frac{J_w}{SRF})$$ (6.3)

where J_n is the joint set number
J_r is the joint roughness number
J_a is the joint alteration number
J_w is the joint water reduction factor
SRF is the stress reduction factor.

Q values are set on a logarithmic scale in a range of 0.001 to 1000. Similar to the process of RMR value selection, the values of individual factors for Q are chosen on the basis of measurements and observations of the field conditions. Similarly, the Q system has gone through many years of updating and refinement. The values from the handbook by the Norwegian Geotechnical Institute in NGI (2013) will be used in this textbook.

The six factors in Eqn. 6.3 are grouped in the following three categories:

RQD/J_n – the block size,
J_r/J_a – the interlock shear strength, and
J_w/SRF – the effect of active stress.
Each of them has a specific meaning, to be explained below.

A). Block size

RQD/J_n represents the structure of the rock mass, a crude measure of the block size. RQD is from core logging as described before. J_n is a number to reflect the frequency of joints and has a value from 0.5 (with no or few joints) to 20 (many joints), as given in Table 6.11. The quotient therefore has two extreme values calculated with the extreme values of RQD and J_n:

Maximum RQD/minimum $J_n = 100/0.5 = 200$

Minimum RQD/maximum $J_n = 10/20 = 0.5$.

The results give a ratio of 400:1. This can be considered approximately as a measure of the different block/particle sizes. In calculation, the percent value of RQD is used.

Table 6.11 Joint set number J_n in Q-system (After NGI 2013).

1. Jointing condition	J_n	Notes
A. Massive, no or few joints	0.5-1.0	
B. One joint set	2	
C. One joint set plus random	3	1) for intersections,
D. 2 joint sets	4	use 3 x J_n.
E. 2 joint sets plus random joints	6	
F. 3 joint sets	9	2) for portals,
G. 3 joint sets plus random joints	12	use 2 x J_n.
H. ≥ 4 joint sets, random heavily jointed "sugar cube"	15	
J. Crushed rock, earth-like.	20	

B). Interlock shear strength

The quotient J_r/J_a represents the roughness and frictional characteristics of the joint walls and filling materials. Higher values are given to rough, unaltered joints in direct contact since such surfaces are expected to show the highest shear strength of joints. They will also dilate strongly when sheared, which is especially favourable to the ground stability around excavations. When rock joints have thin clay mineral coatings and fillings, the strength is reduced significantly. However, rock wall contact after small shear movement has occurred may be a very important factor for preserving the excavation from ultimate failure. When there is no rock wall contact, the condition is very unfavourable to the ground stability. Individual values for J_r and J_a are selected from Tables 6.12a and 6.12b, respectively.

Some of the terminology in Table 6.12 are explained and demonstrated in the following pages.

Table. 6.12a Joint roughness number J_r (After NGI 2013).

2. Joint Surface Condition	J_r	Notes
a. Rock wall contact and		
		1). description refers to small scale and intermediate scale features in that order
b. Rock wall contact before 10 cm shear movement		
A. Discontinuous joints	4	
B. Rough or irregular, undulating	3	2). add 1.0 if the mean spacing of the relevant joint set is greater than 3 m (dependent on opening size)
C. Smooth, undulating	2	
D. Slickensided, undulating	1.5	
E. Rough, irregular, planar	1.5	
F. Smooth, planar	1.0	3). $J_r = 0.5$ can be used for planar, slickensided joints having lineations, provided that the lineations are oriented in the estimated sliding direction.
G. Slickensided, planar	0.5	
c. No rock wall contact when sheared		
H. Zone containing clay minerals thick enough to prevent rock wall contact when sheared	1.0	

Table. 6.12b Joint alternation number J_a (After NGI 2013).

3. Joint Alteration	J_a	ϕ_r (°)
a. Rock wall contact *(no mineral fillings, only coatings)*		
A. Tightly healed, hard, non-softening, impermeable filling, i.e. quartz or epidote	0.75	
B. Unaltered joint walls, surface staining only	1.0	25-35
C. Slightly altered joint walls. Non-softening mineral coatings; sandy particles, clay-free disintegrated rock	2	25-30
D. Silty or sandy clay coatings, small clay fraction (non-softening).	3	20-25
E. Softening or low friction clay mineral coatings, i.e., kaolinite or mica. Also chlorite, talc gypsum, graphite, etc. and small quantities of swelling clays.	4	8-16
b. Rock wall contact before 10 cm shear *(thin mineral fillings)*		
F. Sandy particles, clay-free disintegrated rock, etc.	4	25-30
G. Strongly over-consolidated, non-softening clay mineral fillings (continuous but <5mm thick).	6	16-24
H. Medium or low over-consolidation, softening, clay mineral fillings (continuous but <5mm thick).	8	12-16
J. Swelling-clay fillings, i.e., montmorillonite (continuous but <5mm thick). J_a value depends on % of swelling clay-size particles.	8-12	6-12
c. No rock wall contact when sheared *(thick mineral fillings)*		
K. Zones or bands of disintegrated or crushed rock. Strongly over-consolidated.	6	16-24
L. Zones or bands of clay, disintegrated or crushed rock. Medium or low over-consolidation or softening fillings.	8	12-16
M. Zones or bands of clay, disintegrated or crushed rock. Swelling clay. J_a value depends on % of swelling clay-size particles.	8-12	6-12
N. Thick continuous zones or bands of clay. Strongly over-consolidated.	10	12-16
O. Thick, continuous zones or bands of clay. Medium to low over-consolidation.	13	12-16
P. Thick, continuous zones or bands with clay. Swelling clay. J_a value depends on % of swelling clay-size particles.	13-20	6-12

Note: Values of ϕ_r, the residual friction angle, are intended as an approximate guideline to the mineral properties of the alteration products, if present.

Note a). Illustration of joint surface condition for J_r

- A, discontinuous
  joints

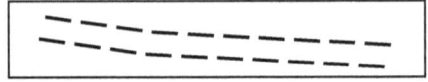

- B to D, irregular or
  undulating joints,

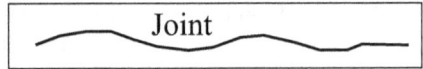

Joint

- Roughness increases
  as the arrow
  indicates.

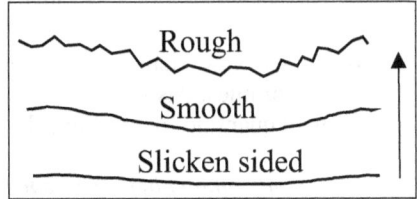

Rough

Smooth

Slicken sided

- E to G, planar joints

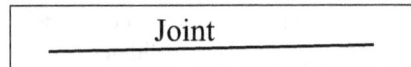

Joint

- H, joints with soft fillings

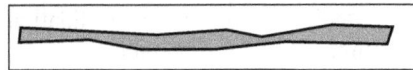

Note b). elaboration on joint alteration for J_a
- A to B, tight, no filling joints,
- C to E, with some soft fillings and surface alteration,
- F to J, with soft fillings but rock walls are expected to be in contact after 10 cm shear movement.
- K to P, with thick fillings, or a weak zone and no rock wall contacts are expected.

  From A ---> P, the degree of alteration increases.
Both J_r and J_a should refer to the joint set which is most likely to initiate failure.

C). Effect of active stress

The quotient J_w/SRF includes two stress parameters to be selected from Table 6.13a and 6.13b, respectively. J_w is a measure of water pressure which has an adverse effect on the shear strength because it reduces the effective normal stress. SRF (Stress Reduction Factor) is a measure of loosening load in excavation through shear zones, rock stress in competent rock and squeezing load in plastic-like incompetent rocks.

Table. 6.13a Joint water reduction number (After NGI 2013).

4. Joint Water Reduction	J_w
A. Dry excavations or minor inflow (humid or a few drips)	1.0
B. Medium inflow, occasional outwash of joint fillings (many drips / "rain")	0.66
C. Jet inflow or high pressure in competent rock with unfilled joints	0.5
D. Large inflow or high pressure, considerable outwash of joint fillings	0.33
E. Exceptionally high inflow or water pressure decaying with time. Causes outwash of material and may cave in.	0.2-0.1
F. Exceptionally high inflow or water pressure continuing without noticeable decay. Causes outwash of material and may cave in.	0.1-0.05

Note: 1) Factors C to F are crude estimates. Increase J_w if rock is drained or grouting is done.
2) Special problems caused by ice formation are not considered.

Table. 6.13b Stress reduction factor (After NGI 2013).

5. Stress Reduction Factor	SRF
a. Weak zones intersecting underground opening, which may cause loosening of rock mass	
A. Multiple occurrences of weak zones within a short section containing clay or chemically disintegrated, very loose surrounding rock (any depth), or long sections with incompetent (weak) rock (any depth). For squeezing, see M and N	10
B. Multiple shear zones within a short section in competent clay-free rock with loose surrounding rock (any depth)	7.5
C. Single weak zones with or without clay or chemical disintegrated rock (depth ≤ 50 m)	5
D. Loose, open joints, heavily jointed or "sugar cube", etc. (any depth)	5
E. Single weak zones with or without clay or chemical disintegrated rock (depth > 50 m)	2.5

Note: 1) Reduce these values of SRF by 25-50% if the weak zones only influence but do not intersect the underground opening.

Table. 6.13b (Continued)

5. Stress Reduction Factor			SRF
b. Competent, mainly massive rock, stress problems	σ_c/σ_1	σ_θ/σ_c	SRF
F. Low stress, near surface, open joints	>200	<0.01	2.5
G. Medium stress, favourable stress condition	200-10	0.01-0.3	1.0
H. High stress, very tight structure. Usually favourable to stability. (May also be unfavourable to stability dependent on the orientation of stresses compared to jointing / weakness planes*.)	10-5	0.3-0.4	0.5-2 (2-5*)
J. Moderate spalling and/or slabbing after > 1hour in massive rock.	5-3	0.5-0.65	5-50
K. Spalling or rock burst after a few minutes in massive rock.	3-2	0.65-1	50-200
L. Heavy rock burst and immediate dynamic deformation in massive rock.	<2	>1	200-400

Note: 2) For strongly anisotropic virgin stress field, if $5 \leq \sigma_1/\sigma_3 \leq 10$, reduce σ_c to $0.75\sigma_c$. If $\sigma_1/\sigma_3 > 10$, reduce σ_c to $0.5\sigma_c$.
3) σ_θ is the maximum tangential stress.
4) If the depth of the crown below the surface is less than the span, increase SRF from 2.5 to 5.

c. Squeezing rock: plastic deformation in incompetent rock under the influence of high pressure	σ_θ/σ_c	SRF
M. Mild squeezing rock pressure	1-5	5-10
N. Heavy squeezing rock pressure	>5	10-20

Note: 5) Determination of squeezing rock conditions must be made according to relevant literature (i.e., Singh et al. 1992 and Grimstad 1996).

d. Swelling rock: chemical swelling activity depending on the presence of water	SRF
O. Mild swelling rock pressure	5-10
P. Heavy swelling rock pressure	10-15

Notes: in Table 6.13b, for SRF,
- A to E, weak zones and joints with thick soft fillings, or open joints (D)

- F to L, competent rock, problem mainly from stress, not joints.
- M to N, squeezing ground, soft rock,
- O to P, swelling ground, water sensitive rock.

For complete descriptions, please refer to Barton et al. (1974) and NGI (2013). In addition to the above factors, there will undoubtedly be other factors that could be added to the Q-system. The joint orientation, for example, seems to be important as shown in the RMR system but is not included in the Q-system. There are other factors, which may not be so important in comparison to those in the system.

Application of the Q index:

The value of Q can be used to determine the requirement of supports for underground excavation and to select ground support, which will be discussed in more detail in the next chapter. The following formula determines the maximum dimension (span or height) allowed in a type of rock mass:

$$D_{max} = D_e \, ESR \hspace{3cm} (6.4)$$

where D_{max} is the allowed maximum dimension on the cross section of an opening as shown in Fig. 6.3, D_e is the equivalent dimension, and ESR is the Excavation Support Ratio.

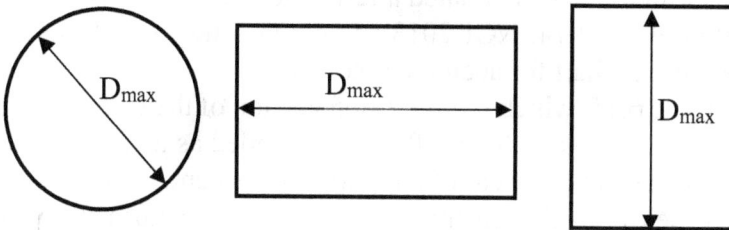

Fig. 6.3 Maximum opening dimension.

ESR may be considered as the inverse of the factor of safety. The values of ESR range from 0.8 (for long-term use) to 5 (for short-term or temporary use), as shown in Table 6.14.

Table 6.14 Excavation category and ESR values.

A. Temporary mine openings	3 to 5
B. Permanent mine openings, water tunnels, pilot tunnels, drifts and headings for large excavations.	1.6
C. Storage rooms, water treatment plants, minor road and railway tunnels, surge chambers, access tunnels.	1.3
D. Power stations, major road and railway tunnels, civil defence chambers, portals, intersections.	1.0
E. Underground nuclear power stations, railway stations, sports and public facilities, factories.	0.8

Equivalent dimension D_e is defined as

$$D_e = D_{max} \text{ (maximum span or height) / ESR} \qquad (6.4a)$$

D_e is to be determined based on the Q value using the graphic Q-chart (NGI 2013), which was developed based on many case studies and has been updated a few times over many years (e.g., Barton el al. 1974, NGI 2013). Users are suggested to use the most recent chart for accurate assessment.

Table 6.15, which is a modified version of the Q-chart based on the handbook of NGI (2013), is intended as a reference and approximate assessment of support requirement. An example is given below to demonstrate how to use the values in the Q-system and Table 6.15.

Table 6.15. Approximate estimate of equivalent dimension De and support requirements based on Q value (After NGI 2013).

Q	Rock Quality	De≤3 (m)	≤5 (m)	≤10 (m)	≤20 (m)	≤50 (m)	>50 (m)
≤0.004	Exceptionally	6/7*	7	8	8/9	9	9
≤ 0.01	poor	6/7	7	7	8	9	9
≤ 0.04	Extremely	5/6	6/7	7	7/8	8/9	9
≤ 0.1	poor	4/5	5/6	6/7	7	8/9	9
≤ 0.4	Very	1/4	4/5	5/6	6/7	7/9	9
≤ 1	poor	1/3	4	4/5	5/6	6/7/9	9
≤ 4	Poor	1	3	3/4	4/5	5/6	9
≤ 10	Fair	1	1	3	3/4	4/5	5/9
≤ 40	Good	1	1	1/3	3	3/4/5	4/5/9
≤ 100	Very good	1	1	1	2	2/3	3/9
≤ 400	Extremely good	1	1	1	1/2	2	2/9

* type of support defined below:

1 Unsupported or spot bolting
2 Spot bolting
3 Systematic bolting + Sfr (5-6 cm)
4 Systematic bolting + Sfr (6-9 cm)
5 Systematic bolting + Sfr (9-12 cm)

6 Systematic bolting + Sfr (12 - 15 cm) + RR 1
7 Systematic bolting + Sfr (15 - 25 cm) + RR 2
8 Cast concrete lining, or Systematic bolting + Sfr (≥25 cm) + RR 3
9 Special evaluation required.

Sfr - Fibre reinforced shotcrete, **RR** - reinforced ribs, 3 types.
Bolting spacing in types 1 to 4: ≤1.0 m (if Q = 0.1 - 4) ~ 4.0 m (if Q ≥ 100),
Bolting spacing in types 5 to 9: ≤1.0 m (if Q ≤0.01) ~ 2.5 m (if Q = 20).

Example 6.3

An underground subway is to be excavated in a sandstone at a depth of 60 m. There are two sets of joints with some randomly oriented joints all around. The surfaces of the major joints are relatively smooth. The joints are planar with some impermeable fillings. The RQD value is 65% and there is some water seepage.

Questions:

    a) What is the maximum allowable span without support?
    b) If the subway is designed for a 10 m span, what type of support is required?

Solutions:

a) Based on the provided information, the selected ratings and the results are shown below:

Item	Description/Value	Table #		Rating
RQD	65%	-		65
Joint number	2	6.11.1E	J_n	6
Joint roughness	smooth, planar	6.12a.2F	J_r	1.0
Joint alteration	impermeable filling	6.12b.3A	J_a	0.75
Water factor	seepage	6.13a.4A	J_w	1.0
Stress factor	shallow, stress issue	6.13b.5F	SRF	2.5

$\therefore \quad Q = (65/6) \times (1/0.75) \times (1/2.5) = 5.8$

Determine the maximum allowable unsupported span:

    If the Q-chart from NGI (2013) is used, draw a vertical line upwards at Q = 5.8 intersecting the first inclined line, go horizontally to the left, meeting the vertical axis at D_e = 4.2.

    The subway is a major civil tunnel, in Category D, the ESR is 1.0 (Table 6.14). The maximum unsupported span by Eqn. 6.4

$$D_{max} = ESR \times D_e = 4.2 \text{ m}.$$

If Table 6.15 is used as an estimate, with no or spot bolting as in note (1), at Q = 5.8, the D_e is expected between 3 and 5 m. Even at Q = 10, a 5 m span still needs bolting support. Therefore, D_e will be approximately 4 m.

b). When the designed span is 10 m,

$$D_e = \text{span}/ ESR = 10/1.0 = 10.$$

If the original Q-chart is used, draw a vertical line at Q = 5.8 upwards and a horizontal line at D_e = 10 to the right, intersecting in zone (3). The suggested supports in that zone are systematic bolting with 5 – 6 cm fibre reinforced shotcrete, in order to maintain the ground stability for the design.

If Table 6.15 is used, the estimated support is also systematic bolting (note 3).

6.5 Correlation and Application of RMR and Q Systems

As shown before, the two systems are quite similar in some ways but with different emphasis:

- both systems assign numerical ratings to quantify field factors and are useful in making a difficult practical decision,
- the RMR system includes joint orientation but ignores active stress conditions,
- the Q system includes stress condition but not joint orientation, which is partially considered by J_r & J_a, (orientation is very important for underground mining, especially in deep cases in comparison to shallow tunnels),
- in extremely weak ground (e.g., squeezing, swelling & flowing ground), the RMR system is not recommended, instead, the Q system should be used.

Although there is no direct connection between the two systems, there is an approximate relationship between them, which can be expressed approximately as

$$RMR \approx 9 \ln Q + 44 \qquad (6.5)$$

It is very important to remember that both systems may not be adequate if a strong joint dominates the rock mass behaviour. The joint should then be treated individually.

Estimation of the in-situ deformation modulus

As discussed before, the Young's modulus of rock specimens determined in laboratory tests can not be directly applicable to the field condition, in most cases, because of the difference in scale and presence of fractures, etc. It is therefore important to have some way to estimate the value of rock mass modulus in the field. There have been various attempts to do this. Some approximate empirical relationships between the field modulus, E_m, and the rock mass classification index values have been presented in Hoek et al. (1995). They are summarized below.

Bieniawski's (1989) method:

$$E_m \,(\text{GPa}) = 2\, RMR - 100, \; RMR \in (50, 85) \qquad (6.6)$$

Barton's (1976) method:

$$E_m \,(\text{GPa}) = 25\, Log_{10}Q, \; Q \in (2,110) \qquad (6.7)$$

Serafim and Pereira's (1983) method:

$$E_m \,(\text{GPa}) = 10^{(RMR-10)/40}, \; RMR \in (24, 85) \qquad (6.8)$$

In the above equations, the Q and RMR values are limited to the ranges reported in the original database.

Equations 6.6 and 6.8 are also illustrated in Fig. 6.4. When rock mass classification results are available, the above empirical formulae may be used to estimate the field modulus. However, special attention should be paid to the data range (effective values of RMR or Q) when using one of these empirical formulae. Beyond that range, there may be errors.

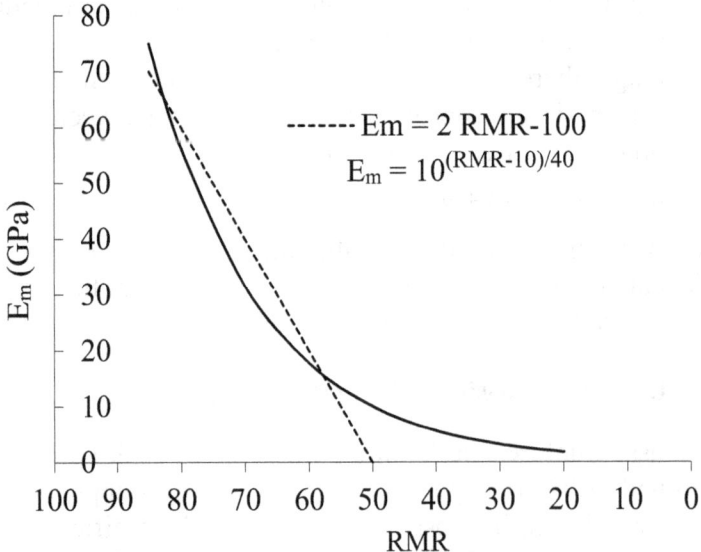

Fig. 6.4 Field modulus calculated by Eqns. 6.6 and 6.8.

When no rock mass classification results are available, an alternative method presented in Hoek et al. (1995) may be used to estimate the field modulus based on the rock mass jointing structure and joint surface conditions. The suggested values of E_m are presented graphically in Fig. 6.5.

The details on rock mass jointing structure and joint surface conditions have been described in the previous chapter, with definition given in Table 5.3.

For convenience, the estimated GSI values are also included in Fig. 6.5 for each category of rock mass. As can be seen, Eqn. 6.8 has a very similar pattern to the empirical data shown in Fig. 6.5.

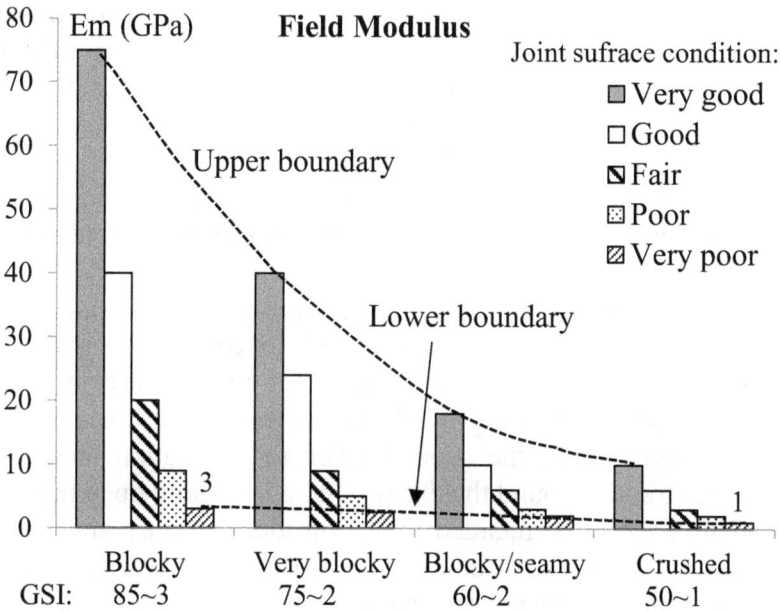

Fig. 6.5 Estimate of field rock mass modulus based on rock mass and joint conditions (After Hoek et al. 1995).

Chapter 7

Stresses in Underground Excavations

σThe stress field, which exists in an undisturbed rock mass is related to the weight of the overlying strata and the geological history in the ground. This stress field is disturbed when an underground opening is created. This disturbance causes stress concentration, inducing a new state of stress in the vicinity. In some cases, the induced stresses are high enough to exceed the strength of the rock mass, causing rock failure. Failure of the rock adjacent to the excavation boundary can lead to instability, which may take place in a form of gradual closure of the excavated opening, roof fall and slabbing of sidewalls, or, in extreme cases, rock burst. Rock burst is explosive rock failure, usually occurring when strong brittle rock is subject to high stress.

In order to understand the mechanism of stress-induced instability and the measures required to control the ground stability, it is necessary to understand not only the strength but also the stresses in the ground. This chapter will discuss the stresses existing in the ground prior to excavation and the induced stresses around the excavated underground openings.

This chapter is intended to be introductory only and very much simplified. The discussion is restricted to the use of elastic theory, which means in practical terms that it is limited to use in excavation design in hard rocks. Readers who are concerned with excavations in weak / soft rocks such as salt or potaσ, which have time-dependent properties, are advised to consult publications dealing specifically with the behaviour of these types of material. Nevertheless, the concept and content discussed here will provide practical engineers with a general understanding of what is happening in the field once an opening is excavated.

7.1 In-situ Stresses

Prior to excavations, the rock in the earth's crust is subject to a pre-existing three-dimensional stress field, an in-situ state of stresses. These stresses must be measured or estimated before we can calculate the stress re-distribution around any man-made openings. Numerous publications on field stresses are available.

At a depth of Z below the surface, the state of stresses is in three dimensions, as illustrated in Fig. 7.1, due to the weight of overburden, tectonic stress and geological residual stress.

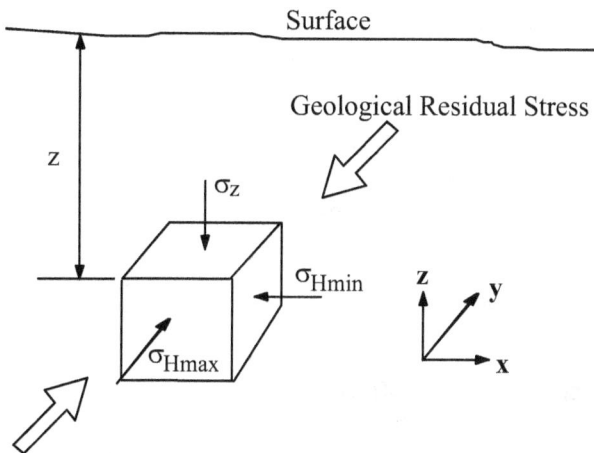

Fig. 7.1 Three-dimensional stress state underground.

As discussed in Chapter 4, a three-dimensional stress field has six independent stress components: three normal stresses in three mutually orthogonal directions (e.g., x, y and z) and three shear stresses. There are three principal stresses as well, which are also perpendicular to each other. In many cases, one principal stress is approximately vertical in the field. Thus, the three major components of the stress field are the vertical, the maximum and the minimum horizontal stresses, coinciding approximately with the principal stresses.

Thus, to determine the state of a three-dimensional stress field, we need to know the magnitude of all three major stresses and the direction of the two major stresses on the horizontal plane. It is possible to measure the magnitudes and the directions of the in-situ stresses using stress measurement methods, but it is often very expensive. Alternatively, they can often be estimated by other means.

1). Vertical stress

The vertical stress in the ground is known to be proportional to the average rock density and the depth, Z, below the surface,

$$\sigma_v = \gamma Z \tag{7.1}$$

where Z is given in (m), σ_v in (MPa) and γ is the unit weight of rock and is usually expressed in terms of (MPa/m depth) for convenience (1 KN/m^3 = 10^{-3} MPa/m). σ_v may also be labelled as σ_z as in Fig. 7.1 when z is chosen to be vertical.

Results of field measurements in a number of countries compiled by Hoek and Brown (1980a) are shown in Figs. 7.2 and 7.3. The vertical stress has an average unit weight of

$$\gamma_{av} = 0.027 \text{ MPa/m depth} \tag{7.2}$$

and varies (±13MPa) between the upper and lower boundaries.

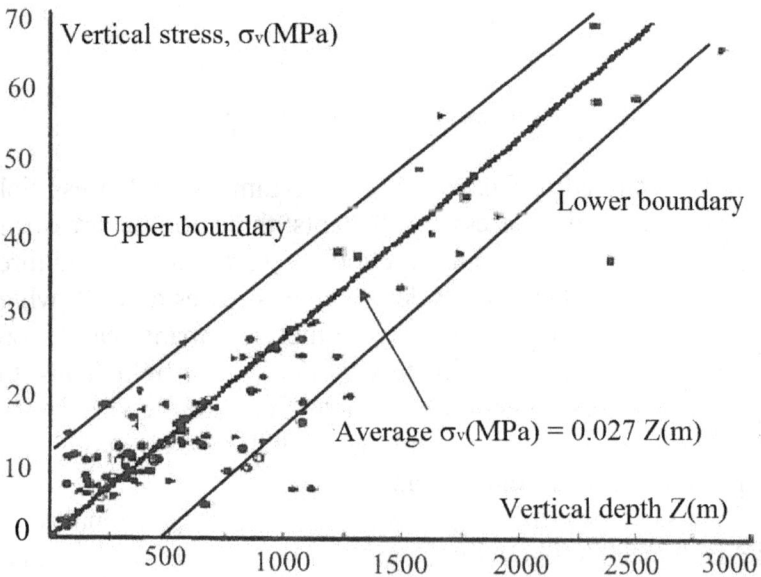

Fig. 7.2 Vertical stress below surface (after Hoek & Brown 1980a).

2). Horizontal stress to vertical stress ratio

The average ratio of the measured horizontal stress to vertical stress varies widely in the upper 1000 m depth (Fig. 7.3), from a range of 1.5 - 3.5 near the surface, to a range of 0.5 - 2.0 at 1000 m depth. This variation becomes smaller as the depth increases. Based on the trend, this ratio is expected to converge to a hydrostatic stress condition of unity below 3000 m depth. The general conclusion is that the horizontal stress in many parts of the world is higher than the vertical stress, i.e.,

$\sigma_H > \sigma_v$, at depth down to 2500 m in most cases.

At depth < 500 m, the horizontal stress is significantly higher than the vertical stress, up to 2.5 times more.

The average stress ratio is defined as

$$k_1 = \sigma_{H\text{-av}} / \sigma_v \qquad\qquad (7.3)$$

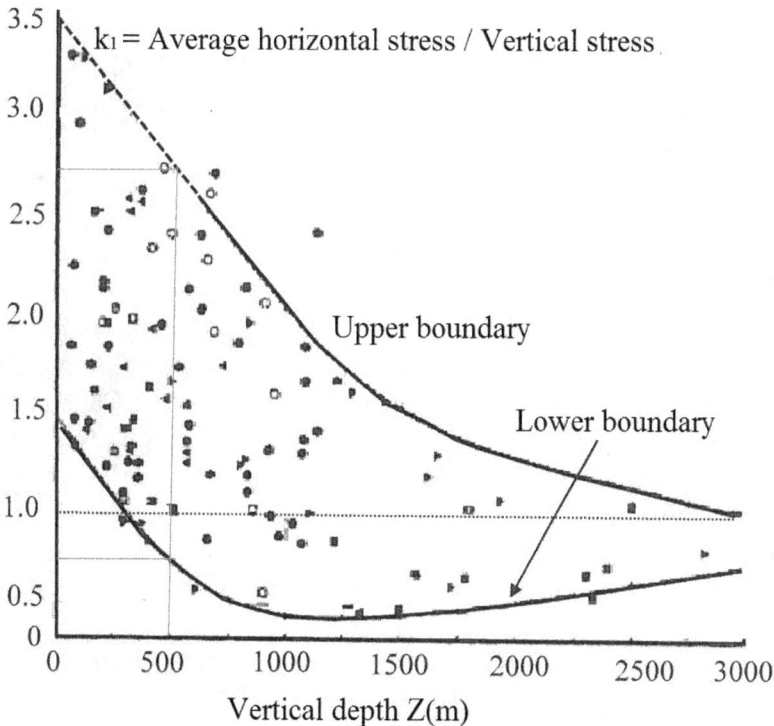

Fig. 7.3 Horizontal stress to vertical stress ratio below surface (After Hoek and Brown 1980a).

The ratio k_1 has a wide range of variation. Because of the scattering of the data, it is difficult to express this trend with a simple formula. The best way to estimate the stress ratio k_1 is to interpret in Fig. 7.3, at a particular depth, between the lower and upper boundaries. For example, at 500 m depth, $k_1 = 0.7 \sim 2.7$.

3). Directions of major horizontal stresses

In the field, the stress state is defined by the three principal stresses (σ_1, σ_2 and σ_3). As indicated earlier, the vertical stress is approximately parallel to the smallest principal stress (σ_3). The other two principal stresses are then approximately within the horizontal plane. Unless field measurement is carried out to determine the exact magnitudes and directions, the three major stresses are normally expressed as the vertical stress σ_v, and the major and the minor horizontal stresses (σ_{Hmax} and σ_{Hmin}). The major and minor horizontal stress directions are reported in the World Stress Map database and the results for North America and East Asia (Heidbach et al. 2008) are shown in Fig. 7.4. Readers interested in other areas of the world and recent updates are suggested to visit the latest database (Heidbach et al. 2016).

In Fig. 7.4, the long arrows represent the direction of σ_{Hmax} and the direction perpendicular to them is for σ_{Hmin}. These directions are scattered over continents but clustered in local areas. In general, they can be summarized as follows:
- the major horizontal stress directions vary around the globe,
- they seem to follow a strong trend in local regions, for example, along the Rocky Mountains in western Canada,
- the trend seems to coincide with the boundaries between the continental plates, or major fault zones in the earth's crust, usually perpendicular to these geological structures. Typical examples are the California Fault zone on the west coast of the United States and the Himalaya Fault zone of western China.

The scattering of arrows in Fig. 7.4 may be a result of various sources, such as, local geological and topographic sometimes possible measurement errors, etc. However, this topic will not be discussed any further in this textbook.

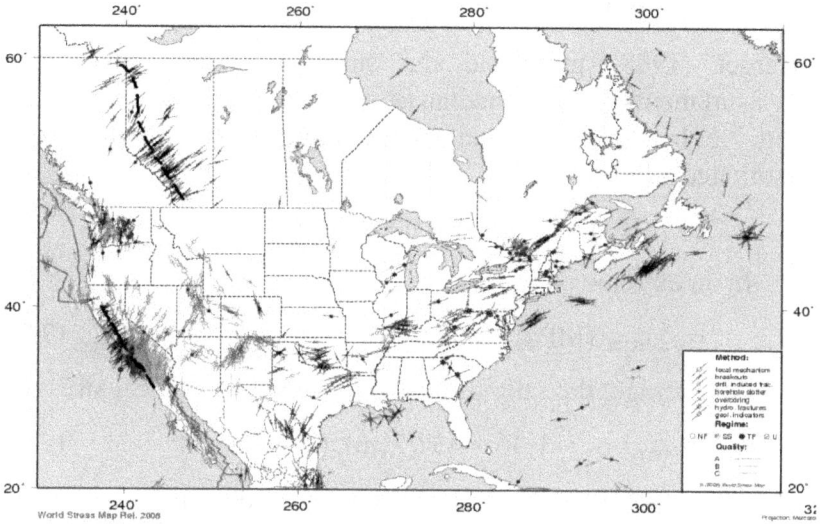

a)   Major horizontal stress direction in North America

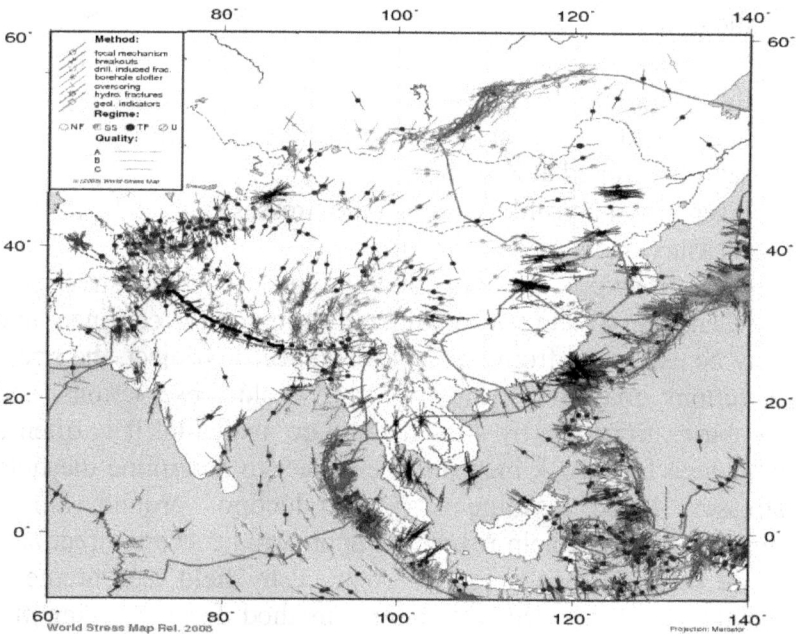

b) Major horizontal stress direction in Asia

Fig. 7.4 Sample major horizontal stress, (From Heidbach et al. 2008).

4). Stresses in the Canadian shield

Herget (1988) published the statistical stress data from measurements in the Canadian Shield, which are summarized in Fig. 7.5. In general, vertical stress in the Canadian Shield can be estimated as

$$\sigma_v \ (MPa) = (0.026 \pm 0.006) \ Z(m) \qquad (7.4)$$

In an extreme case,

$$\sigma_{v\text{-extreme}} \ (MPa) = 0.0602 \ Z(m) \qquad (7.5)$$

On average, the ratio of horizontal stress to vertical stress is

$$\sigma_{Hmax} / \sigma_v = 1.46 + 357/Z(m), \ (Z \geq 100 \ m) \qquad (7.6)$$

$$\sigma_{Hmin} / \sigma_v = 1.10 + 167/Z(m), \ (Z \geq 100 \ m) \qquad (7.7)$$

Note: For the stresses in the Canadian Shield:
- at depth $Z \leq 2500$ m, $\sigma_H > \sigma_v$,
- the ratio σ_H/σ_v varies widely at depth ≤ 500 m, e.g., from 1.4 to 2.2 at $Z = 500$ m, and from 2.5 to 4.5 at $Z = 100$ m,
- the stress ratio by Eqns. 7.6 and 7.7 seems excessively high at $Z < 100$ m. It is suggested that the maximum value of the ratio be capped at $Z = 100$ m,
- it seems not possible to predict the stresses at great depth from measurements made near surface.

The directions of the major and minor horizontal stresses in the Canadian Shield vary with the location on the continent and may be estimated from Fig. 7.4. In a specific project, the stress directions may be estimated from field observations and databases from nearby sites if they are available. It is often a challenge for a rock mechanics engineer to determine the field stresses. Once a mine is in production, ground failure observations will help reveal the approximate stress directions. More accurate results can be obtained by field measurement. However, until recently the only method for measuring the complete 3D stresses was by overcoring (Vreede 1981), though an alternative method based on differential-direction drilling (D^3) and borehole convergence has recently been developed by the author and his graduate student Dr. Lin (Lin 2019).

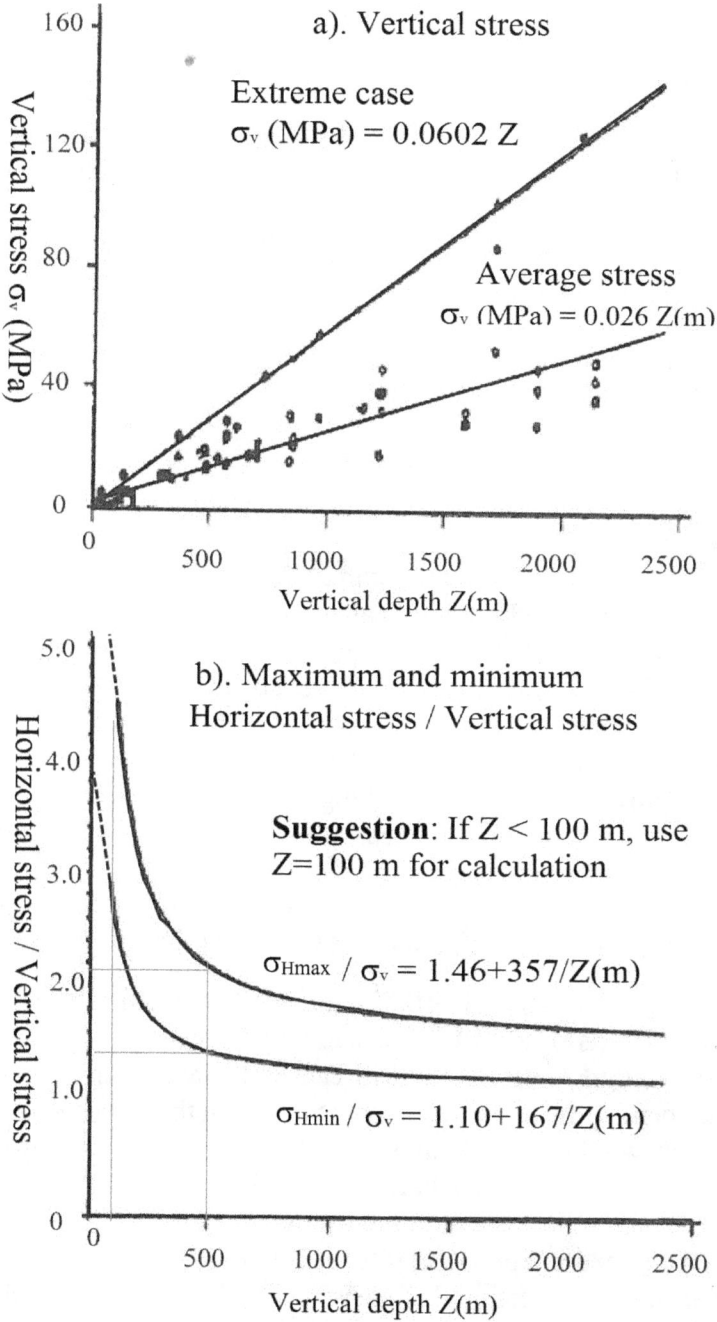

Fig. 7.5 Stresses in Canadian Shield (Modified after Herget 1988).

7.2 Stresses in a Cross Section of a Drift

Stresses at a point in the rock mass

In order to fully define the state of stress at a point within a rock mass, it is necessary to consider an infinitesimal cubic volume enclosing the point in question, as shown in Fig. 4.9. To completely define the stress state at a point, we need to know the six independent stress components. The choice of a volume element with faces parallel to the x, y and z co-ordinate axes was however completely arbitrary. In practice, for convenience it may be necessary to choose an element with faces perpendicular to a set of local axes inclined to the global axes (East, North and Vertical). For example, the element can be oriented such that one pair of faces is parallel to a structural feature such as a joint plane, on which the stresses are to be calculated.

Calculation of stresses in three dimensions is not a simple task. However, the burden of calculation involved in studying a three-dimensional stress problem can often be reduced by considering the two-dimensional stress distribution in one of the principal planes, such as the horizontal plane or a vertical plane through a principal horizontal stress direction. Even when it is not totally justified in making this simplification, two-dimensional stress analyses can provide a useful guide to the nature of three-dimensional stress distributions.

Plane strain condition in a cross section of a drift

For most underground drifts, haul ways, etc., the dimension along the axis of a drift is substantially longer than that in the cross section. A very long drift can be treated as an infinitely long opening. In this case, no deformation in the direction along the drift would exist. A simple explanation behind the logic is that if there were axial deformation, the drift would become longer or shorter. In reality, that is not what occurs.

In engineering practice, if the length of a drift is at least five times the cross-sectional dimension, a plane strain condition exists in central segment. In this case, plane strain theory as discussed in Chapter 4 applies. Stress distribution around a drift will be assumed to be the same for all cross sections, except near the ends of the drift and unless there is some change in local geology and ground condition. In the following sections, our discussion will focus on plane strain conditions.

Example 7.1

A mine is to be designed in Northern Canada at 1000 m depth. The rock mass consists of mostly Greywacke with some weak joints. The orebody dips 80° with a dip direction of 120° measured on the horizontal plane from the North. The major horizontal stress is assumed to be parallel to the orebody. Estimate the in-situ stresses on the cross section of
  a) a level drift driven along the orebody,
  b) a cross cut, which is perpendicular to the drift.

Solutions:

Based on the information provided, first draw a diagram to show the orebody, drift, cross cut and the in-situ stresses, with the drift // to the orebody and the cross cut ⊥ to the orebody, as shown on the right.

Since σ_{Hmax} is // to the orebody, the σ_{Hmin} must be ⊥ to the orebody. The three major field stresses are as shown in the diagram.

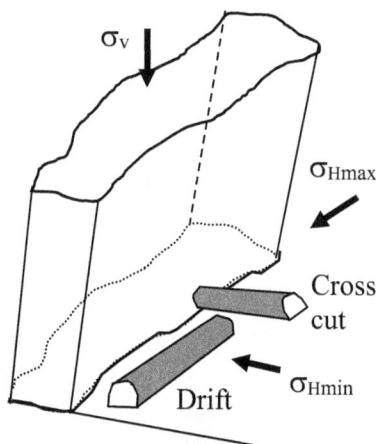

Use Eqn. 7.4 to calculate the average vertical stress (use average unit weight unless specified):

$$\sigma_v = 0.026\ Z = 0.026\ (MPa/m) \times 1000\ (m) = 26\ MPa$$

Use Eqns. 7. 6 and 7.7 to calculate the horizontal stresses:

$$\sigma_{Hmax} = (1.46 + 357/Z)\ \sigma_v$$
$$= (1.46 + 357/1000) \times 26 = 47.2\ MPa$$

$$\sigma_{Hmin} = (1.10 + 167/Z)\,\sigma_v$$
$$= (1.1 + 167/1000) \times 26 = 32.9 \text{ MPa}$$

a). On the cross section of the drift, the two stress components are, as shown:

$\sigma_{Hmin} = 32.9$ MPa $(= \sigma_{max})$

$\sigma_v = 26$ MPa $(= \sigma_{min})$

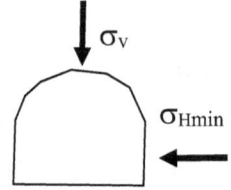

b). On the cross section of the cross cut, the two stress components are:

$\sigma_{Hmax} = 47.2$ MPa $(= \sigma_{max})$

$\sigma_v = 26$ MPa $(= \sigma_{min})$

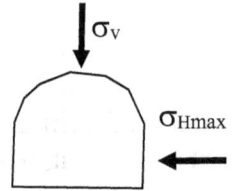

7.3 Stress Distribution around a Single Opening

Stresses existing in the ground are not something we can see. Excavation of an opening in the rock mass disturbs the original stress field and stresses re-distribute around the excavation until a new stress equilibrium is reached, or the rock fails.

To demonstrate stress distribution, let us look at the following examples:

a). Air flow

Imagine that on a very windy day, if you stand beside a building facing the wind direction, you will feel a stronger wind than in an open place. If you move to the opposite side of the building, you may feel little or no wind at all. That is the effect of air flow redistribution due to obstruction of the building.

If there are two buildings side by side and close to each other, when you stand between the buildings facing the wind, you will feel a much stronger wind and you may not be able to stand stably. That is the effect of two buildings obstructing the wind.

If you stand in different locations around the building, you may feel differently. In these examples, you felt the changes of airflow.

b). Water flow

Imagine that you stand on a river bank where water flows gently in the river. If there is a rock or a tree stump poking out of the middle of the river, you will see a change in the pattern of water flow: it by-passes the object and then rejoins afterwards as shown in Fig. 7.6. The directions of the streamlines are changed around the object. In this example, you saw the change of water flow.

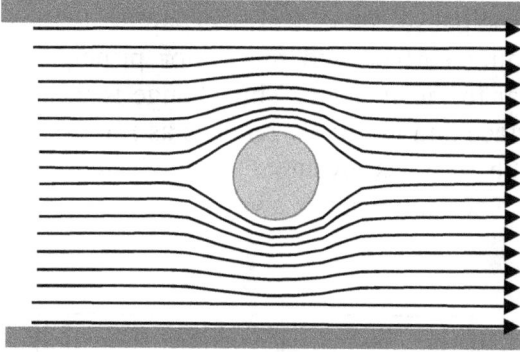

Fig. 7.6 Change of streamline of water flow due to obstruction.

In the case of underground stresses, we cannot see or feel the stress changes caused by excavation of an opening. From Fig. 7.6, a comparison can be made between the water flow and stress field:

- streamline crowding on both sides of the object resembles stress concentration in the stress field on both sides of an opening. Since the water flow rate should be the same in all cross sections along the river of the same width, a crowding area means water flowing faster. In relation to the stress, this would mean higher stress values or a stress concentration.

- streamline separation (i.e., the empty space before or after the object) resembles stress relaxation in the stress field next to the opening.

In a stress field, the stresses are transmitted from one point to the next point through adjacent solid material particles. Excavation of an opening in the rock mass removes that part of the solid materials and changes the characteristics. Stresses cannot be transmitted through a void. Similar to the air flow pattern, stresses have to be transmitted through adjacent remaining materials. The final effect is stress concentration on both sides of the opening.

In a two-dimensional stress field, there are two principal stresses (σ_1, σ_2). Their magnitudes and directions at every point around an opening are changed after excavation. Figure 7.7 illustrates the change patterns of the principal stresses after excavation of an opening. On or near the surface of the opening, the major principal stresses are always parallel to the tangent line on the opening surface and the minor principal stresses are perpendicular to the surface. This change is the largest on the opening surface and gradually reduces as it gets away from the opening. The stresses eventually match the original stress condition at a certain distance away.

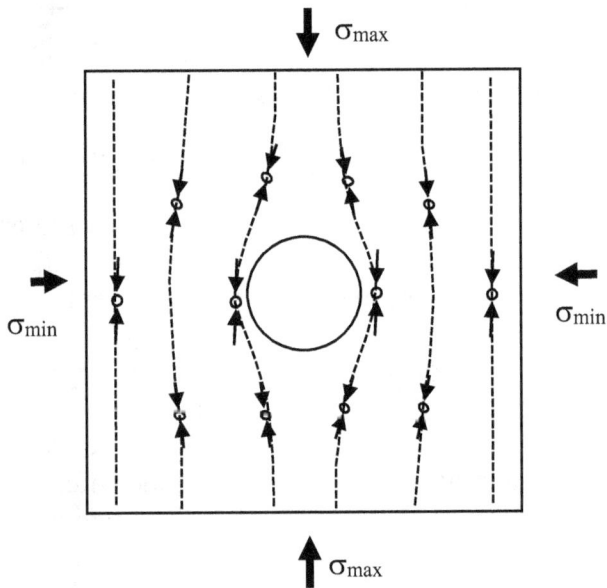

Fig. 7.7 Changes of principal stress directions due to an opening (After Hoek and Brown 1980a).

For a two-dimensional stress field, the one dimensional effect shown in Fig. 7.6 will overlap each other in two perpendicular directions. Figure 7.7 shows the final results of the effects from stresses in both directions. The remaining question now is how to determine the stress changes and the new principal stresses in the disturbed area.

Stress change and distribution around a circular opening

Stress changes around an opening are very complex and vary with the shape of the opening and the location. We will start with the simplest case of a circular opening, for which analytical solutions are available from the theory of elasticity.

In a two-dimensional stress field defined by (σ_{max}, σ_{min}) as shown in Fig. 7.8, if a circular opening is excavated, from the elasticity theory based on isotropic and linear elastic assumptions, the new stresses at any point around the opening are functions of the original stresses and its location relative to the opening. More detail is given in Jaeger and Cook (1976).

In a polar coordinate (r, θ) system, the new stresses can be calculated by the following equations.

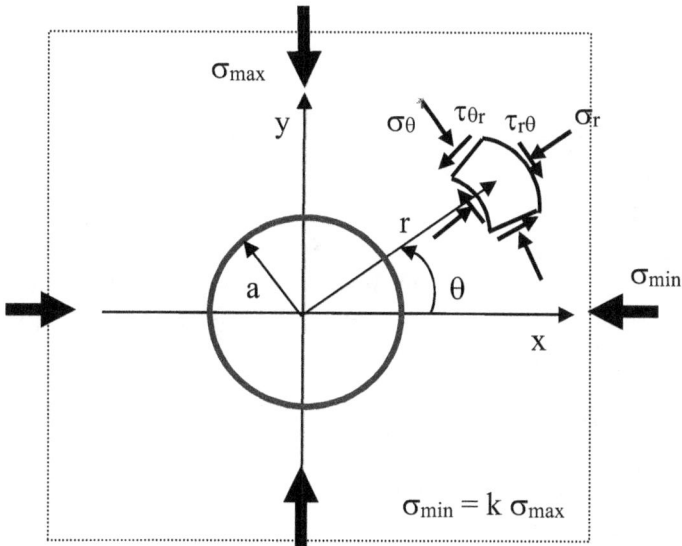

Fig. 7.8 New stresses around a circular opening. (**Note**: The angle θ is positive if measured counter-clockwise from the minimum stress direction.)

$$\sigma_r = 0.5\sigma_{max}[(1+k)(1-\frac{a^2}{r^2}) - (1-k)(1-4\frac{a^2}{r^2}+3\frac{a^4}{r^4})\cos 2\theta] \quad (7.8a)$$

$$\sigma_\theta = 0.5\sigma_{max}[(1+k)(1+\frac{a^2}{r^2}) + (1-k)(1+3\frac{a^4}{r^4})\cos 2\theta] \quad (7.8b)$$

$$\tau_{r\theta} = 0.5\sigma_{max}[-(1-k)(1+2\frac{a^2}{r^2}-3\frac{a^4}{r^4})\sin 2\theta] \quad (7.8c)$$

where θ is measured from the direction of σ_{min} (i.e., x axis) counter-clockwise (i.e., the positive direction) and the field stress ratio k is defined as

$$k = \sigma_{min} / \sigma_{max} \quad (7.9)$$

where $k \leq 1.0$.

From Eqns. 4.10 and 4.8, respectively, the new principal stresses and their directions can be determined using the following equations.

$$\sigma_{1,2} = \frac{\sigma_r + \sigma_\theta}{2} \pm \sqrt{(\frac{\sigma_r - \sigma_\theta}{2})^2 + \tau_{r\theta}^2} \quad (7.10a)$$

$$\tan 2\theta_a = \frac{\tau_{r\theta}}{(\sigma_r - \sigma_\theta)/2} \quad (7.10b)$$

where the angle θ_a is the direction of σ_1 or σ_2 at that point, measured from the reference axis r, counter-clockwise (i.e., the positive direction).

The immediate vicinity of the excavated area is of most concern for its stability. Failure usually starts in this area and damage may show up first on the opening surface. Therefore, stress conditions on the opening surface will be of great interest.

Stresses on the opening surface

The opening surface is the boundary between the solid rock mass and the newly created void. In Eqn. 7.8, r = a at any location on the opening surface. Therefore,

$\sigma_r = 0$ and $\tau_{r\theta} = 0$,

$$\sigma_\theta = \sigma_{max}[(1+k) + 2(1-k)\cos 2\theta] \quad (7.11)$$

The tangential normal stress σ_θ is the only stress component on the opening boundary. It parallels the tangent line on the boundary and varies in magnitude with angle θ. As $\cos 2\theta$ is a periodic function and its value varies between ±1.0, the value of σ_θ will vary between its maximum and minimum, which define the most critical points on the boundary.

As demonstrated in Fig. 7.9, at Points A and B, along the direction of the minimum field stress σ_{min}, $\theta = 0°$ and $180°$, respectively (at the mid height on the side walls for a horizontal opening) and the tangential stress reaches its maximum

$$\sigma_{\theta max} = \sigma_{max} (3 - k) \tag{7.12}$$

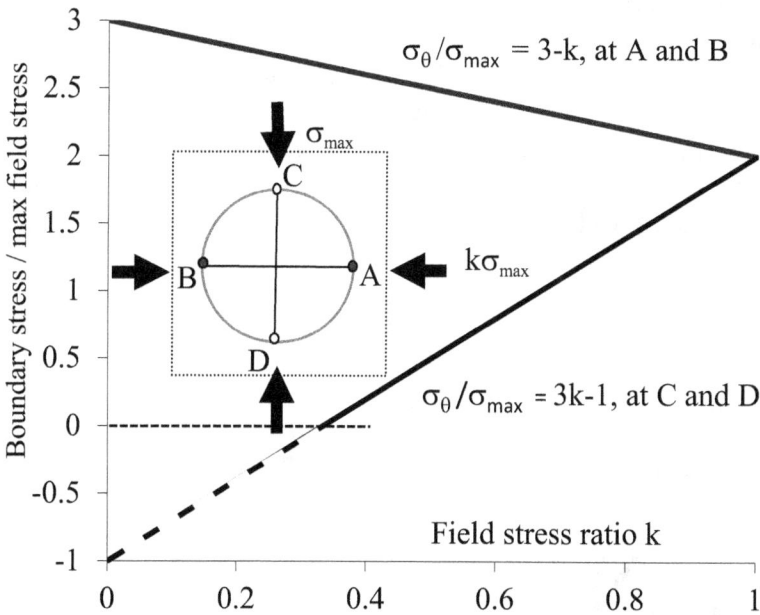

Fig. 7.9 Change of boundary stress at critical locations with the in-situ stress ratio k.

Equation 7.12 means $\sigma_{\theta max}$ is always in compression. The possible failure mode in this condition will be compression or shear as defined by the Mohr's failure envelop described in Chapter 5.

At Points C and D, along the direction of the maximum field stress σ_{max}, $\theta = \pm 90°$, respectively (C and D will be at the roof and the floor for a horizontal opening), the tangential stress reaches its minimum value:

$$\sigma_{\theta min} = \sigma_{max}(3k - 1) \qquad (7.13)$$

When $k < 1/3$, $\sigma_{\theta min} < 0$ and tension develops at C and D, as shown in Fig. 7.9. Then, possible tensile failure develops since rock mass has very low tensile strength.

Distance of stress disturbance

Equation 7.8 indicates that the new stresses vary with a/r, the opening radius to distance ratio. The stress change caused by excavation is the largest on the opening boundary (r = a). It decreases with distance from the opening and eventually vanishes. Figure 7.10 illustrates the stress change with distance along the r direction, parallel to the minimum field stress. At Point A (r/a = 1.0), the change is the largest with the highest stress concentration.

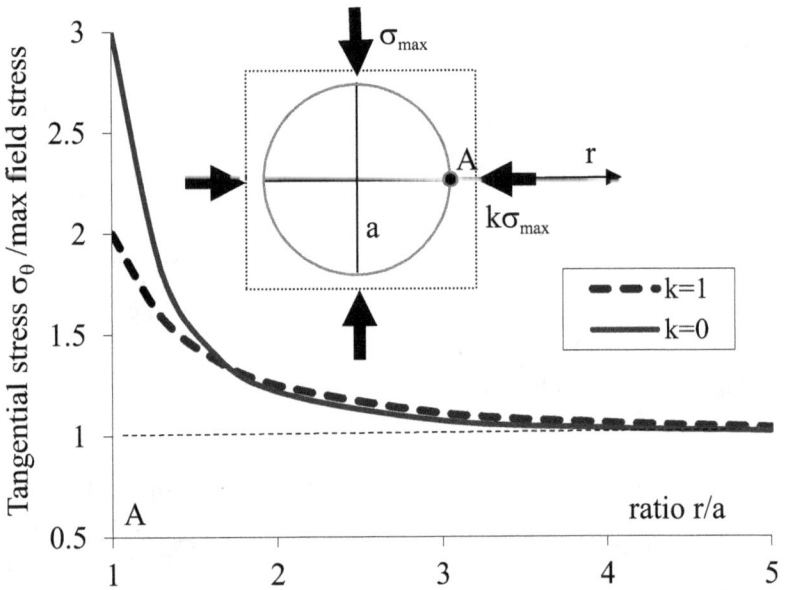

Fig. 7.10 Stress changes with radial distance along the r axis through A.

Let us look at the mid-height horizontal line through point A as an example. On this line, $\theta = 0°$. By Eqn. 7.8, $\tau_{r\theta} = 0$ and $\sigma_r < \sigma_\theta$. The ratio of $\sigma_\theta / \sigma_{max}$ varying with the distance ratio r/a is plotted in Fig. 7.10 for the two extreme cases of k =1 and k=0.

As shown, most stress changes take place within a zone of three times the opening dimension, i.e., $1.0 \leq r/a \leq 3.0$. At a distance where r/a = 3 ~ 5, the change becomes small to very small and when $r/a \geq 5$, the change is negligible. The largest stress change takes place on the opening boundary (r = a). The stress concentration is 2 when k = 1 (the hydro-static condition) and 3 when k = 0 (one-dimensional loading).

It is important to point out that the induced stresses
- depend on the field stresses, the relative position and the opening shape,
- are independent from material properties for linear elastic rock,
- and independent from the opening size.

However, the ground stability will be affected by the size of an opening because the stability depends on the stresses, rock mass properties and the presence of any discontinuities. A larger opening is most likely to intersect more weak joints and in turn reduce the overall strength of the rock mass.

Due to stress redistribution after excavation, the principal stresses at every point in the vicinity are also changed both in magnitude and direction. In the original stress field, the principal stresses were uniform and pointed to the same directions throughout the area. In the new stress field, the principal stresses are "rerouted" as demonstrated in Fig. 7.7 and their magnitudes and directions are all changed. More examples of principal stress changes and contours due to excavations are available in Hoek and Brown (1980a).

7.4 Influence of Opening Shape and Orientation

In practice in the field, many openings are non-circular. It is therefore necessary to examine the stress distributions around openings of different shapes. As the excavation shape changes, the induced stresses are expected to change accordingly and the critical points on the boundary will change as well.

Elliptical opening

An elliptical opening is defined in Fig. 7.11, subject to field stresses σ_{max} and σ_{min}, where $L_1 \parallel \sigma_{min}$ and $L_2 \parallel \sigma_{max}$. Based on elasticity theory, the tangential stress at critical Point A (or B) along the direction of σ_{min} on the opening boundary is given as

$$\sigma_{\theta A} = \sigma_{max} (1 + 2 L_1/L_2 - k) \tag{7.14}$$

or as $\quad \sigma_{\theta A} = \sigma_{max} (1 + \sqrt{2 L_1/r_A} - k) \tag{7.14a}$

and at Point C (or D) along the direction of σ_{max} is given as

$$\sigma_{\theta C} = \sigma_{max} [k (1 + 2 L_2/L_1) - 1] \tag{7.15}$$

or as $\quad \sigma_{\theta C} = \sigma_{max} [k (1 + \sqrt{2 L_2/r_C}) - 1] \tag{7.15a}$

where r_A and r_C are the radii of curvature at A and C.

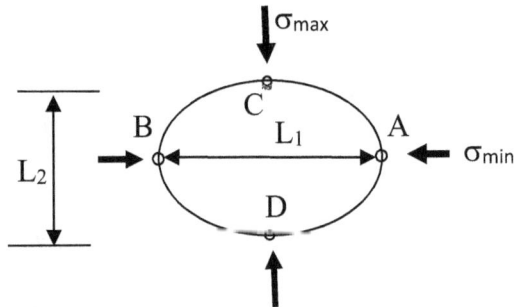

Fig. 7.11 Dimension of elliptical opening.

The above equations have in fact included Eqns. 7.12 and 7.13 for a circular opening, which can be considered as a special case where $L_1 = L_2$. Equations 7.14 and 7.15 include two important factors: the curvature at a corner, represented by radius r and the opening orientation, (represented by L_2/L_1 ratio), relative to the major stress in the field. The effects of the two factors are examined below.

Effect of the opening orientation

To examine the effect of the opening orientation, let us first consider one-dimensional loading (i.e., k = 0) as shown in Fig. 7.12, where the opening is orientated in different directions relative to the field stress σ_{max}.

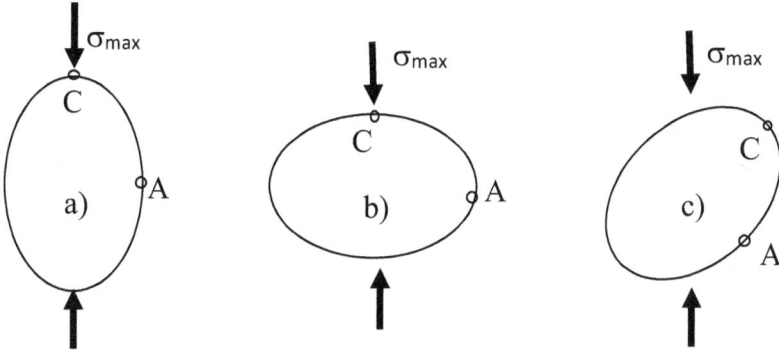

Fig. 7.12 Critical boundary stress around elliptical openings.

For a vertical elliptical opening (on the left in Fig. 7.12) with height/width (H/W) ratio = 2, the tangential stresses at points A (perpendicular to the direction of σ_{max}) and C (along the direction of σ_{max}) are, respectively,

$$\sigma_{\theta A} = 2\sigma_{max} \text{ and } \sigma_{\theta C} = -\sigma_{max}, \quad (H = 2W) \quad (7.16)$$

For a horizontal elliptical opening (in the middle), the two stresses are

$$\sigma_{\theta A} = 5\sigma_{max} \text{ and } \sigma_{\theta C} = -\sigma_{max}, \quad (H = 0.5W) \quad (7.17)$$

It can be seen that the stress concentration at A has increased from 2 to 5 when the longer axis of the opening is changed from being parallel to being perpendicular to the stress σ_{max}. When the opening is inclined from the field stress σ_{max}, the stress concentration will vary between those in the above two cases.

For the linear elastic condition, the effect of two-dimensional loading shown in Fig. 7.11 can be determined by superimposing the effects of one-dimensional loading of Fig. 7.12 upon each other. This will partially reduce the above effect.

For example, if another stress, $\sigma_{min} = k \, \sigma_{max}$, is applied to Fig. 7.12, the values in Eqns. 7.16 and 7.17 will be reduced by an amount equivalent to that caused by σ_{min}.

Therefore, to reduce stress concentration and the risk of stress-related ground instability, the longer axis of an opening should be // to the direction of the maximum field stress.

Effect of corner curvature

For rectangular and other irregularly shaped openings, there are no analytical solutions available and stress analysis will rely on numerical modelling. However, the observations in elliptical opening as shown in Eqns. 7.14a and 7.15a are still applicable.

As shown in the two equations, the influence of the radius of curvature on the stress concentrations at the corners of the excavations are significant. The larger the radius of curvature, the lower the compressive stress concentration is, and vice versa. The most favourable stress condition is achieved at rounded corners of an opening when the radius of curvature takes the maximum possible value. This corresponds to an ovaloidal opening, where the corner radius is half the height of a flat-lying opening (or half the width of an upright-standing opening), as shown on the right in Fig. 7.13. Therefore, for optimum ground stability, a corner should have the maximum possible radius.

Design principles for underground openings:
- avoid sharp corners,
- an opening should be oriented with its longer axis // to the maximum field stress in the cross section,
- if the field stress ratio $k = 1$ (hydro-static condition), the optimum opening shape is a circle,
- if $k \neq 1$, a corner should have the largest possible radius.
- if $k \leq 1/3$, tensile stress may occur, be aware of tensile failure.

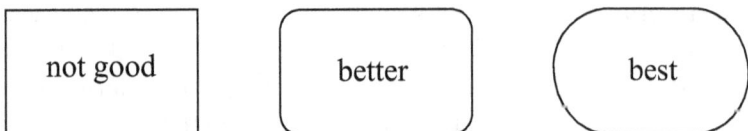

Fig. 7.13 Lowest boundary stresses around an ovaloidal opening.

7.5 Stress Distribution around Multiple Openings

Overlap of stress concentrations

The previous discussions indicated that for a single opening, the stress disturbance zone is three to five times the opening dimension. If two openings are placed outside of their disturbance zones, no influence exists on each other. If both openings are within their disturbance zones, their effects will overlap each other, as illustrated in Fig. 7.14. In the overlapping zone, the shorter the distance, L, between the openings, the higher the stress concentration will be.

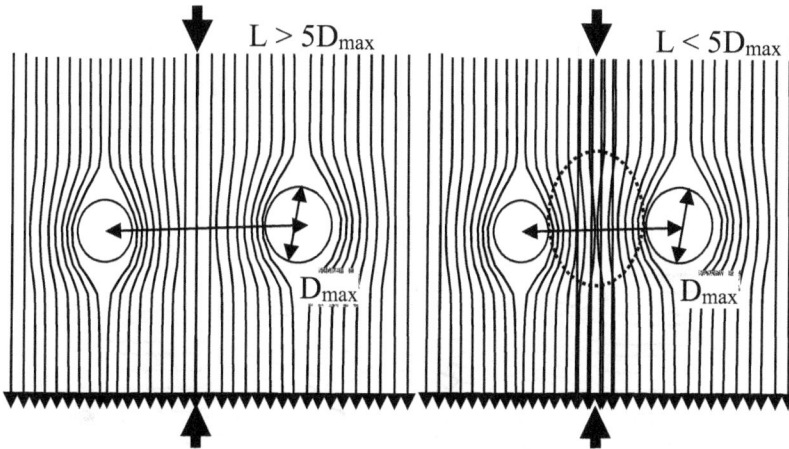

Fig. 7.14 Stress distribution around multiple openings.

In a mine operation, many openings are created for the purpose of ore extraction, resulting in pillars between openings. The stress in a pillar will depend on the width of the pillar relative to the opening size and the height to width ratio of the pillar. In general, the stress in a pillar can be simplified to consist of two components: the average pillar stress and the stress concentration. The stress concentration in a pillar may spread over the whole cross section of the pillar or only near the edges, depending on the pillar size. More discussion on determination of pillar stresses for various types of pillars and pillar design will be given in next chapter.

Stress relaxation zone

As discussed before, the existence of an opening will create a stress disturbance zone in the rock mass. In this zone, some part of the rock mass suffers from higher stress due to stress concentration, while the other part may carry less stress than before. This is indicated by the "empty" or less crowded space in Figs. 7.7 and 7.14, a phenomenon called stress relaxation.

If multiple openings align in parallel along the major field stress direction, the pillars between them may carry lower stress under the protection of the stress relaxation zone (Fig. 7.15). This occurs only where the openings are close to each other, or within the disturbance zone.

This effect is best understood if you stand between two buildings on a very windy day where one building is behind the another, facing the wind. You may feel little or no wind at all.

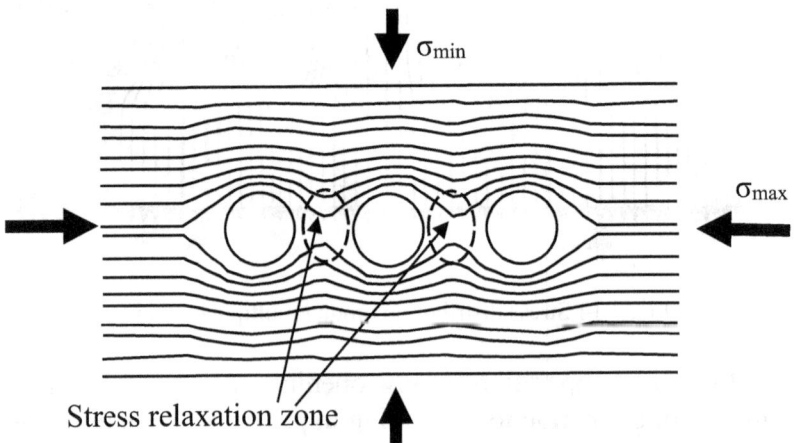

Fig. 7.15 Potential stress relaxation.

Complex 3D pillar problems

While the foregoing discussions are very useful in demonstrating the concept of stress changes and concentrations caused by underground excavations, in many cases, stress issues in the field of a mining operation are never that simple. Excavation of mine openings in various locations and arrangements create three-dimensional stress issues. Fig. 7.16 demonstrates a few typical examples at the intersections between drifts and cross cuts, between draw point and cross cuts, and the stope itself. There are no analytical solutions to these problems. They are normally solved using sophisticated 3D numerical modelling tools. Numerical modelling itself is a separate subject and a brief introduction will be given in a later chapter.

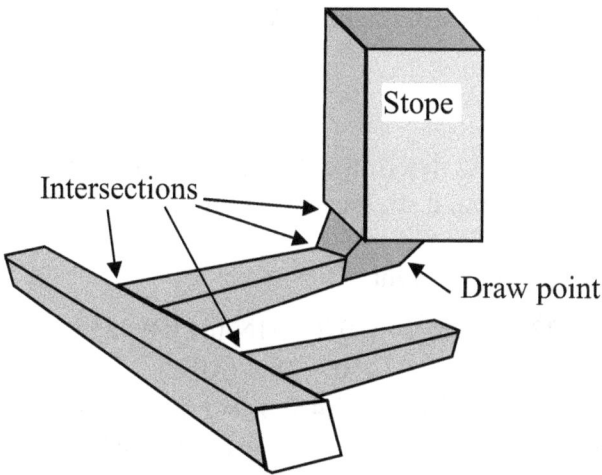

Fig. 7.16 Complex 3D stress distribution at intersections.

Example 7.2

A mine is to be designed in Northern Ontario at a depth of 1450 m, where the average rock mass unit weight is estimated at 2700 kg/m^3.

a). If the maximum horizontal stress in the field σ_{Hmax} is oriented in E-W, what is the best orientation of a main haulage drift if other conditions permit?

b). Based on the choice in a), for a circular drift with radius $r_0 = 3$ m, what are the principle stresses and their directions after excavation on a horizontal line through the center of the drift at distance $r = 2\ r_0$?

c). What are the maximum and minimum stresses on the drift boundary? their locations? their directions?

d). If the drift is non-circular, compare the following choices and choose the best opening shape with rationale.

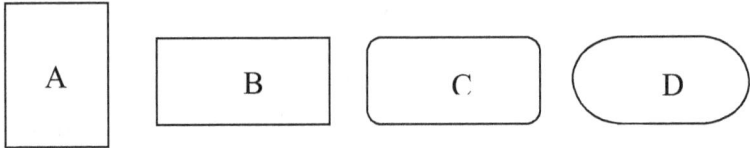

e). If three drifts are running in parallel, is the orientation chosen in a) still the best? Why?

Solutions:

The first step is to determine the vertical, the maximum and the minimum horizontal stresses based on the data in Fig. 7.5:

Based on the unit weight,

$$\gamma = 2700 \text{ kg/m}^3 = 0.0265 \text{ MN/m}^3 = 0.0265 \text{ MPa/m depth},$$
$$\therefore \sigma_v = 0.0265 \times 1450 = 38.4 \text{ (MPa)}$$
$$\sigma_{Hmax} = (357/1450 + 1.46)\,\sigma_v = 1.7\,\sigma_v = 65.3 \text{ (MPa)}$$
$$\sigma_{Hmin} = (167/1450 + 1.1)\,\sigma_v = 1.2\,\sigma_v = 46 \text{ (MPa)}$$

a). A long horizontal drift should align with the maximum horizontal stress in the field to avoid excessively high stress on its cross section. Therefore, the drift should be // to σ_{Hmax}, i.e., in the E-W direction as shown below.

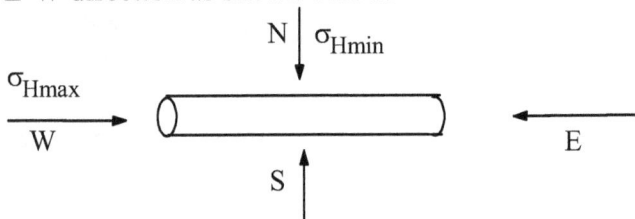

b). To determine the principal stresses at any location around an opening, the first step is to calculate the new stresses (σ_r, σ_θ, $\tau_{r\theta}$,) at that location after excavation, using Eqn. 7.8. The in-situ stresses on the cross section of the drift are σ_v and σ_{hmin} as shown in the margin diagram. The stress ratio is

$k = \sigma_v / \sigma_{Hmin} = 38.4/46 = 0.83$.

The specified location in question "on a horizontal line through the center of the drift at distance $r = 2\ r_0$" is Point A, where $r_0/r = 0.5$ and $\theta = -90°$ (measured from σ_v). Then the new stresses are calculated using Eqn. 7.8, as follows:

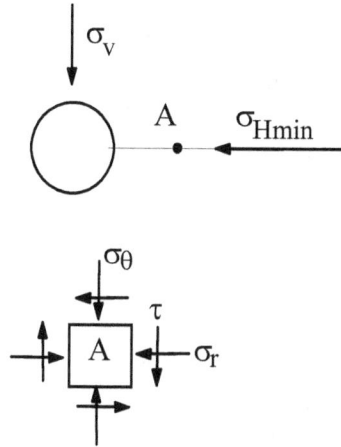

$\sigma_r = 0.5\ \sigma_{Hmin}[(1+k)(1-0.5^2) -$
   $(1-k)(1-4\times0.5^2+3\times0.5^4)\cos2\text{ x }(-90°)]$
   $= 0.5$
   $\times 46\ [(1+0.83)\times0.75 - (1-0.83)\times0.1875\times(-1.0)]$
   $= 32.3\ (\text{MPa})$

$\sigma_\theta = 0.5\ \sigma_{Hmin}[(1+k)(1+0.5^2) +$
   $(1-k)(1+3\times0.5^4)\cos2\text{ x }(-90°)]$
   $= 0.5$
   $\times 46\ [(1+0.83)\times1.25 - (1-0.83)\times1.1875\times(-1.0)]$
   $= 47.8\ (\text{MPa})$

$\tau_{r\theta} = 0$ because $\sin(-2\times90°) = 0$.
$\therefore$ $\sigma_1 = \sigma_\theta = 47.8$ MPa, the vertical stress at A,
   $\sigma_2 = \sigma_r = 32.3$ MPa, the horizontal stress at A.

Note: If $\tau_{r\theta} \neq 0$, Eqn 7.10 or Mohr's diagram must be used to determine the magnitude and direction of σ_1 and σ_2.

c). The field stresses on the cross section of the drift include σ_v and σ_{Hmin} and the stress ratio $k = 0.83$ (from b). On the drift boundary, $r_0/r = 1.0$. The only stress on the boundary is σ_θ. Based on Fig. 7.9 and the loading condition on the cross section, the maximum boundary stress will be at Points C and D, and the minimum boundary stress at Points A and B.

At points A and B, by Eqn. 7.13,

$$\sigma_{\theta min} = \sigma_{Hmin} (3k-1)$$
$$= 46 (3 \times 0.83 - 1)$$
$$= 68.5 \ (MPa).$$

This is the only stress and its direction
is vertical and tangent to the boundary
at these two points.

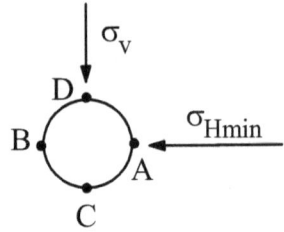

The stresses at Points C and D, by Eqn. 7.12,

$$\sigma_{\theta max} = \sigma_{Hmin} (3-k) = 46 (3 - 0.83) = 99.8 \ (MPa).$$

Again, this is the only stress and its direction is horizontal and
tangent to the boundary at these two points.

d). To determine the best orientation of a drift, we first need
to know the direction of the maximum field stress on the cross
section. In this case, the horizontal field stress is higher than the
vertical stress. For the geometry given below, we conclude

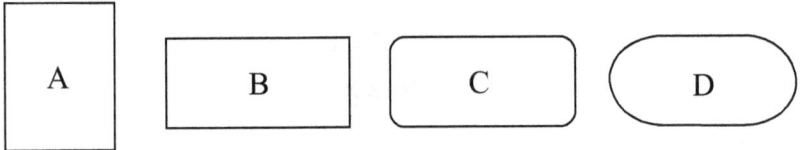

- A: not good, because the long axis is $\perp$ to the maximum
  stress σ_{Hmin} on the cross section,
- B: better than A, but still not as good as C or D because of
  its sharp corners,
- C: better than B because of non-sharp corners,
- D: the best choice because it has the largest radius of
  curvature.

e). For this question, the answer is not so simple anymore.
The three-dimensional stress condition has to be considered.

If the three drifts are far apart (e.g., $> 5 \times 2r_0$), outside of
each other's influence zone, the choice of a) is still valid because
each drift will be in the same stress condition as if they were
single openings, with σ_{Hmin} as the maximum initial stress on the
cross section.

If the three drifts are close to each other, the answer may be different. In this case, the horizontal stress in the pillars between the drifts must also be considered. If they are relatively short, they may be aligned to be // to σ_{Hmin} to take advantage of the relaxation zone (Fig. 7.15) and to reduce the stress in the pillars. Otherwise, it must be examined in more detail. Numerical modelling will be necessary to examine various scenarios and compare the effects before a decision is made.

Note: When any two openings are close to each other, the overlap effect may be considered, for a linear elastic rock mass, by superposition of the new stresses from each opening in the same coordinate system. This may require stress transformation if different coordinates are used.

7.6 Influence of Weak Planes on Stress Distribution around Openings

In general, the influence of discontinuities on stress distribution of underground excavations in a three-dimensional stress field is very complex. This type of problem is usually dealt with by numerical modelling in three dimensions. However, there are a few simple cases where the effects can be examined in a two-dimensional stress condition.

A weak plane or discontinuity usually has very low shear strength and effectively no tensile strength in the direction perpendicular to the plane. The effect of a weak plane may however be negligible or very significant, depending on the position and orientation of the plane. We will once again consider the circular opening as an example, although the principle applies to other shapes.

To evaluate the effect of a weak plane, the possibility of both shear failure along the plane and tensile failure normal to the plane will be examined. The shear strength on a weak plane is estimated by

$$\tau = \sigma_n \tan \phi \tag{7.17}$$

where cohesion is ignored because of its small value.

If the normal stress on the plane $\sigma_n < 0$, tension exists. Separation of the joint becomes possible, a failure condition.

If the shear stress on the plane $\tau \geq \tau_s$, shear failure as slip along the joint plane becomes possible.

When a joint plane is parallel to the radial direction of an opening through the opening center, the shear stress on the plane is in the r (radial) direction at angle θ, given by Eqn. 7.8c

$$\tau_{r\theta} = 0.5\sigma_{max}[-(1-k)(1+2\tfrac{a^2}{r^2}-3\tfrac{a^4}{r^4})\sin 2\,\theta]$$

However, if a joint plane is not along the r direction, the normal and shear stresses on the joint plane need to be determined using the method discussed in Chapter 4 (Eqns. 4.5 and 4.7).

Let us look at a few scenarios below (based on Brady and Brown 2006).

Case 1. A weak plane perpendicular to the maximum field stress passing the opening center

The stress condition for this case is shown in Fig. 7.17. For the half joint plane on the righthand side of the opening, based on the notation given in Fig. 7.8, the angle between the r direction and the reference axis σ_{min}, $\theta = 0°$. Then shear stress $\tau_{r\theta} = 0$, regardless of r.

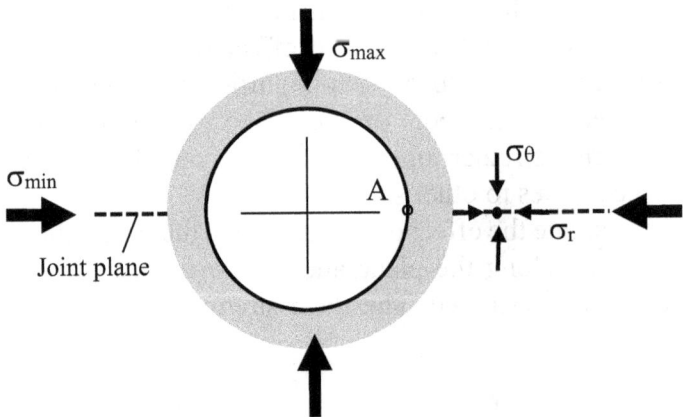

Fig. 7.17 A weak plane through the center and $\perp \sigma_{max}$.

The stresses σ_r and σ_θ are the principal components of compression. The stresses acting on the joint plane are simple:

$\sigma_n = \sigma_\theta > 0$, no separation, and

$\tau = \tau_{r\theta} = 0$, no shear stress.

The same condition exists for the half joint plane on the left side because of the symmetrical stress field. Since the shear stress is zero everywhere on the plane, no slip or shear failure is possible on the joint plane. The only stress at location A, where the joint intersects the opening boundary, is

$$\sigma_\theta = \sigma_{max}(3-k) > 0.$$

Therefore, the presence of the weak plane in this case has no effect on the stress distribution and the opening stability.

Case 2. A weak plane parallel to the maximum field stress passing the opening center

In this case, for the half joint plane on the upper portion (Fig. 7.18), the angle between r direction and the reference axis σ_{min}, $\theta = 90°$.

Then by Eqn. 7.8, $\tau_{r\theta} = 0$, regardless of r. The stresses σ_r and

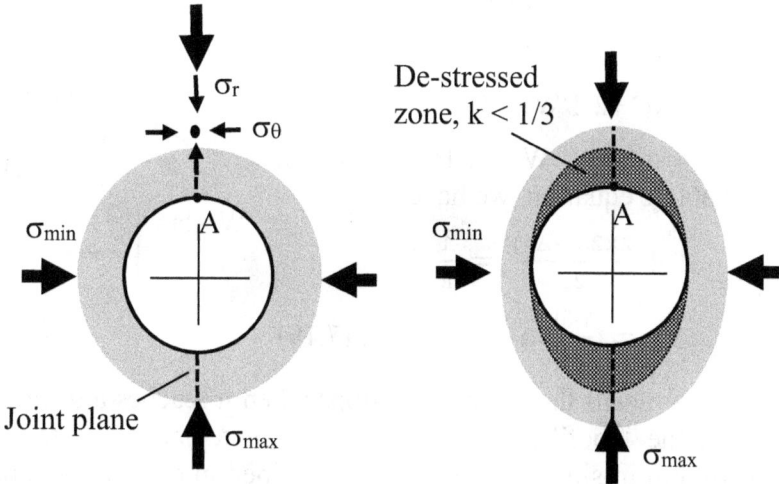

Fig. 7.18 A weak plane through the center and // σ_{max} with possible separation (After Brady and Brown 2006).

σ_θ are then principal components of compression. The stresses acting on the joint plane are

$\sigma_n = \sigma_\theta$ (= σ_{max} (3k - 1) on the opening boundary),

and $\tau = \tau_{r\theta} = 0$, no shear stress.

Again, no shear failure is possible along the joint plane because there is no driving shear force.

However, the stress at the intersection point A on the opening boundary σ_θ is perpendicular to the joint plane and may cause separation if it becomes negative, as shown below

$$\sigma_\theta \begin{cases} > 0, \text{ when k} > 1/3, \text{ in compression} \\ = 0, \text{ when k} = 1/3 \\ < 0, \text{ when k} < 1/3, \text{ in tension} \end{cases} \tag{7.18}$$

Because the joint plane has no tensile strength,
- separation along the plane becomes possible when the in-situ stress ratio k<1/3,
- separation causes partial de-stressing, both in the "roof" and the "floor", forming an elliptical de-stressed zone.

However, the separation process is self-constrained when the height of separation reaches a certain value where $\sigma_\theta = 0$.

By Eqn. 7.15, this would happen when

$$k (1 + 2 L_2/L_1) - 1 = 0$$

i.e., $k(1+2H/W) - 1 = 0$.

Substituting W=2a, H=2(a+Δh) in the above equation, we have

$$1 + \frac{2 \times 2(a+\Delta h)}{2a} = \frac{1}{k}$$

$\therefore$ $\Delta h = \dfrac{1\text{-}3k}{2k} a$ \qquad (7.19)

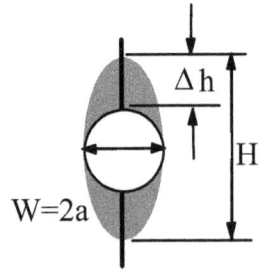

This means that separation stops when it reaches a distance Δh in the "roof" and "floor". If the joint plane is // to the maximum in-situ stress σ_{max} but is inclined in other directions, the same phenomenon will occur along the joint plane.

Case 3. A weak plane perpendicular to the maximum field stress intersecting the opening at angle β

If a horizontal plane intersects the opening boundary at an angle β, as shown in Fig. 7.19, at any point along the plane, there are three stress components (σ_r, σ_θ, $\tau_{r\theta}$). They can be calculated using Eqn. 7.8 based on the location (r, θ) of that point. Then the shear and normal stresses at that location on the joint plane can be calculated using Eqns. 4.5 and 4.7, replacing x by r and y by θ. This is a tedious task and can only be completed using computer software.

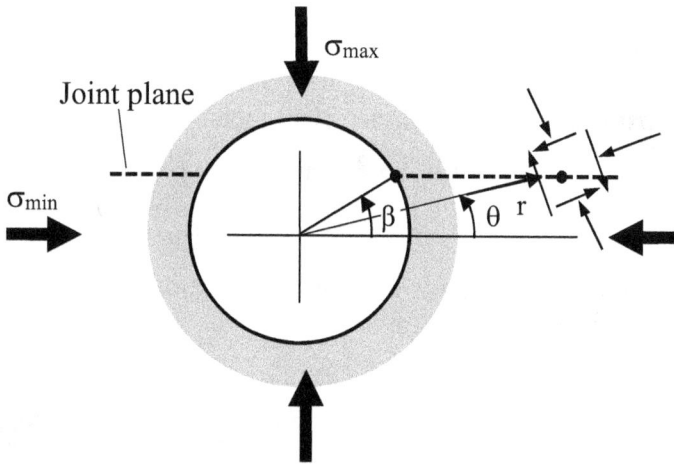

Fig. 7.19 A weak plane intersecting the opening at an angle.

On the opening boundary, however, it will be much simpler since there is only one stress σ_θ. In this case, the tangential stress can be resolved to the normal and shear components as follows:

Consider the r direction as the reference axis. The normal of the joint plane is then defined by angle α.

α = 90 - β.

In Eqns. 4.5 and 4.7, replace θ with α, and substitute (x, y) by (r, θ), we have

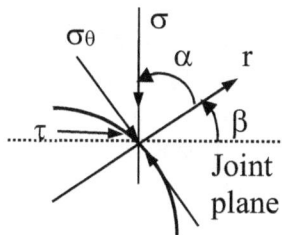

$$\sigma_n = \frac{\sigma_r + \sigma_\theta}{2} + \frac{\sigma_r - \sigma_\theta}{2} \cos 2\alpha + \tau_{r\theta} \sin 2\alpha$$

$$= \sigma_\theta (1 - \cos 2\alpha)/2 \quad = \sigma_\theta \sin^2 \alpha$$

or $\quad \sigma_n = \sigma_\theta \cos^2 \beta$ $\hspace{3cm}$ (7.20)

$$\tau = -\frac{\sigma_r - \sigma_\theta}{2} \sin 2\alpha + \tau_{r\theta} \cos 2\alpha = \sigma_\theta (\sin 2\alpha)/2$$

or $\quad \tau = \sigma_\theta \sin \beta \cos \beta$ $\hspace{3cm}$ (7.21)

With these two stresses, we can now evaluate the shear failure possibility along the plane. From Eqn. 7.17, slip becomes possible if

$$\tau \geq \sigma_n \tan \phi.$$

Substitute Eqns. 7.20 and 7.21 in the above equation,

$$\sigma_\theta \sin \beta \cos \beta \geq \sigma_\theta \cos^2 \beta \tan \phi$$

or $\quad \tan\beta \geq \tan \phi,$

i.e., shear failure may start at the location where the joint intersects the opening at angle

$$\beta \geq \phi$$ $\hspace{3cm}$ (7.22)

Therefore, there is a stable zone defined by 2ϕ. The condition is stable if a plane intersects the opening within the stable zone. If the plane falls in the unstable zone, a de-stressing zone may occur around the intersection points due to spalling.

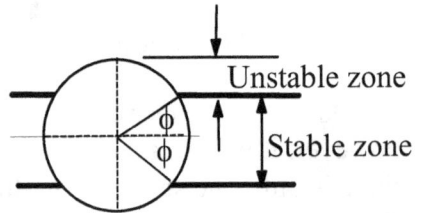

Case 4. An inclined weak plane through the opening center

If an arbitrarily inclined weak plane intersects the opening, the problem is much more complicated. The ultimate effects will depend on the inclination, the location where it intersects the opening and the distance from the opening wall.

To demonstrate the analysis process, let us consider a simple case, where a weak plane inclined at $\beta = 45°$ passes through the centre of the opening as shown in Fig. 7.20. Assume in a given field stress condition, $k = 0.5$. From Eqn. 7.8 we have the normal and shear stresses on the joint plane as

$$\sigma_n = \sigma_\theta = \frac{1.5}{2}\, \sigma_{max}\, \frac{a^2}{r^2} \tag{7.23}$$

$$\tau = -\,\tau_{r\theta} = \frac{0.5}{2}\, \sigma_{max}\, \frac{2a^2}{r^2}\, \frac{3a^4}{r^4} \tag{7.24}$$

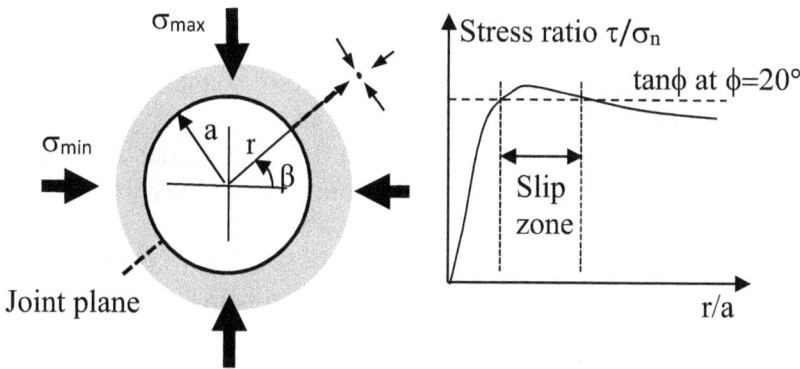

Fig. 7.20 An inclined weak plane intersecting the opening (After Brady and Brown 2006).

The shear failure possibility can be examined by the following comparison:

$$\tau/\sigma_n = \frac{1}{3}\, \frac{1 + 2a^2/r^2 - 3a^4/r^4}{1 + a^2/r^2} \geq \tan \phi$$

The shear to normal stress ratio τ/σ_n can be plotted against the distance ratio r/a. This plot is then compared with the value of $\tan \phi$ (ϕ is the friction angle of the joint plane). The ground becomes unstable if the ratio τ/σ is above the horizontal line defined by $\tan \phi$ as shown in Fig. 7.20.

In this example, the highest point on the curve τ/σ_n is 0.4, corresponding to $\phi =21.8°$. Therefore, for the joint plane with a friction angle of 20°, slip becomes possible in a zone where $\tau/\sigma \geq \tan 20°$ and an area inside the rock mass becomes unstable, as shown in Fig.7.20. An extensive zone of slip may develop along the weak plane in the rock mass.

Case 5. A weak plane perpendicular to the maximum field stress above or below the opening

If a horizontal plane is above or below the opening (Fig. 7.21), not intersecting the opening but still within the stress disturbance zone, it may also have a negative impact on the ground stability. Based on the coordinates (r, θ) of a point on the joint plane, the new stresses $(\sigma_r, \sigma_\theta, \tau_{r\theta})$ can be calculated by Eqns. 7.8 and the normal and shear stresses by Eqns. 4.5 and 4.7.

For the purpose of demonstration, let us consider a simple case where stress ratio $k = 1.0$ and a joint at distance b above the opening (Fig. 7.21).

At any point defined by the distance d on the joint, the normal to the joint plane has an angle α relative to the r direction. The distance r is given by

$$r = \sqrt{b^2 + d^2}.$$

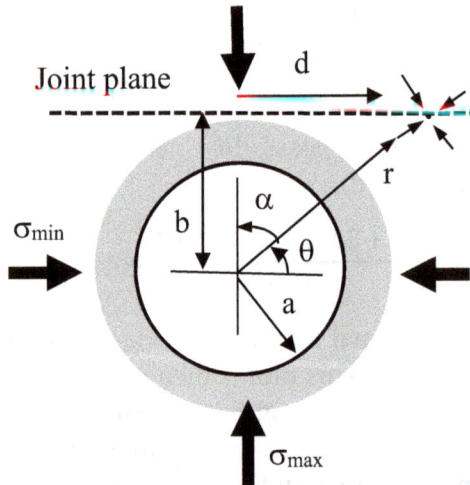

Fig. 7.21 A horizontal weak plane above the opening.

By Eqns. 7.8,

$$\begin{cases} \sigma_r = \sigma_{max} \left(1 - a^2/r^2\right) \\ \sigma_\theta = \sigma_{max} \left(1 + a^2/r^2\right) \\ \tau_{r\theta} = 0 \end{cases} \tag{7.25}$$

The shear and normal stresses on the joint plane are

$$\sigma_n = \tfrac{1}{2}(\sigma_r + \sigma_\theta) + \tfrac{1}{2}(\sigma_r - \sigma_\theta) \cos 2\,\alpha$$

$$= \sigma_{max}\left(1 - \frac{a^2}{r^2} \cos 2\,\alpha\right) \tag{7.26}$$

$$\tau = \tau_{r\theta} \cos 2\,\alpha - \tfrac{1}{2}(\sigma_r - \sigma_\theta) \sin 2\,\alpha$$

$$= \sigma_{max} \frac{a^2}{r^2} \sin 2\,\alpha \tag{7.27}$$

The shear to normal stress ratio is then

$$\tau/\sigma_n = \frac{(a^2/r^2)\,\sin 2\alpha}{1 - (a^2/r^2)\,\cos 2\alpha} \tag{7.28}$$

Again, following the same procedure as shown in Fig. 7.20, the stress ratio τ/σ_n can be plotted against distance d and compared to the $\tan\phi$ value. If the ratio $\tau/\sigma_n \geq \tan\phi$, the condition is unstable and slip may take place. If the ratio τ/σ_n is above the horizontal line defined by $\tan\phi$, there is an unstable zone along the joint plane.

The extent of this zone depends on the location of the plane and the friction angle ϕ. This partial instability may cause stress redistribution and result in damage on the opening surface.

Chapter 8

Stability Consideration in Underground Excavation Design

In all types of rock engineering, underground mining is perhaps the one most concerned with stress distribution and the associated ground stability due to its depth and excavation scale. The day-to-day operation of a mine requires close attention of an experienced rock mechanics or mining engineer. The following discussion addresses some important aspects involved in underground mining and is intended to provide some guidance in design of underground excavations. Properly designed underground infrastructures are expected to reduce unnecessary stress concentrations and to reduce or eliminate some ground instability issues or ground failure possibilities.

8.1 Locations of Major Underground Mine Infrastructures

Primary mine accesses

The primary mine accesses are among the most important mine infrastructures. They include shafts, adits or inclined slopes. They are the only accesses to and from underground and are expected to serve the entire mine for as long as they are needed. The importance of their stability cannot be overly emphasized. These structures are usually well supported during construction. However, proper selection of the location can help reduce the risk of ground instability and the cost of construction and/or support.

Selection of the location of these mine structures will have to take into consideration many factors, such as operation requirements, orebody shape and size, ground condition and

topography, etc. From the point of view of ground stability control, they should be located in an area where:
- the ground is in stable condition and has good rock mass quality
- major geological structures (faults, weak zones, etc.) are avoided, if possible
- the ground stability will not be affected by mine excavations throughout the mine life.

In light of these requirements, a better location is usually in the footwall side of an orebody to avoid any ground instability issues associated with the rock mass failure caused by mine excavations. Rock mass failure happens very frequently and progressively in the hanging wall side in open stope mining. There should be a safe distance from the designed structure to the expected boundary line of ground failure. If for any reason these accesses have to be located in the hanging wall side, the required safe distance will be much greater. In that case, the distance of transportation may become longer for operation and will increase the operation cost of the mine. Figures 8.1 and 8.2 illustrate some examples of proper locations to select and locations to avoid.

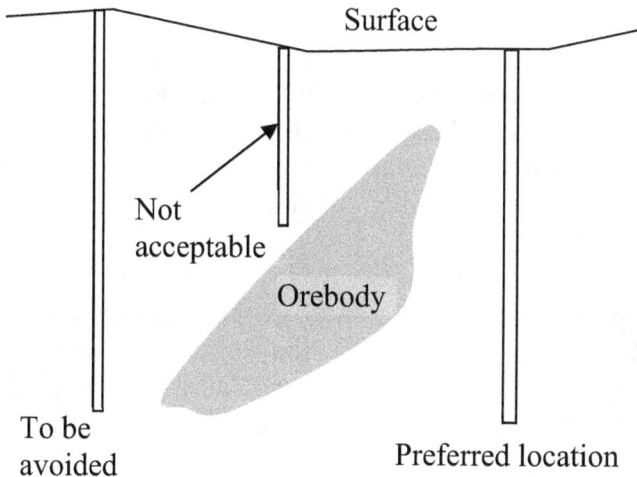

Fig. 8.1 Shaft locations (vertical section view).

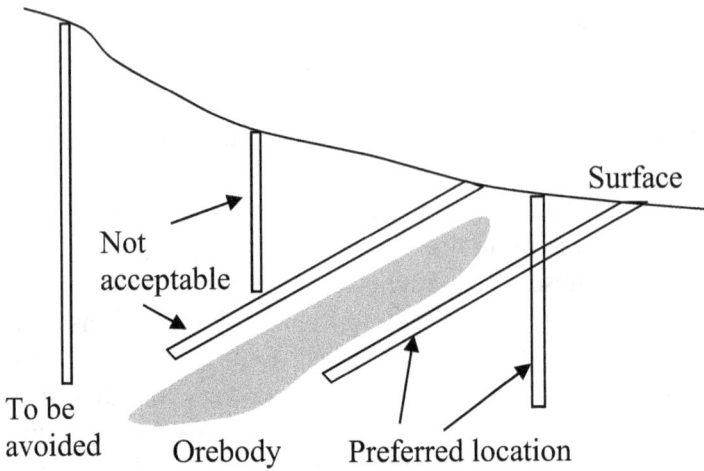

Fig. 8.2 Shaft and slope locations (vertical section view).

Secondary mine accesses

The secondary mine accesses here refer to those serving a major section of a mine, such as, level drifts, panel entries, ventilation raises and ore passes. These mine structures are generally used for the intended section of the mine. Some, however, are often maintained for future use in other parts or a lower level of the mine. The general principles as discussed above still apply.

Level drifts are located at the bottom and some are also at the top of a level. They are usually in the footwall along the orebody and sometimes in the orebody depending on the overall mine design. They should however not be in the hanging wall to avoid ground failure.

Ventilation raises are usually located near the end or at the central part of the orebody, away from the ground failure area, depending on the overall ventilation layout.

Ore passes are mostly in a central location of a production area to be served and in the footwall. Figure 8.3 shows some examples of their locations.

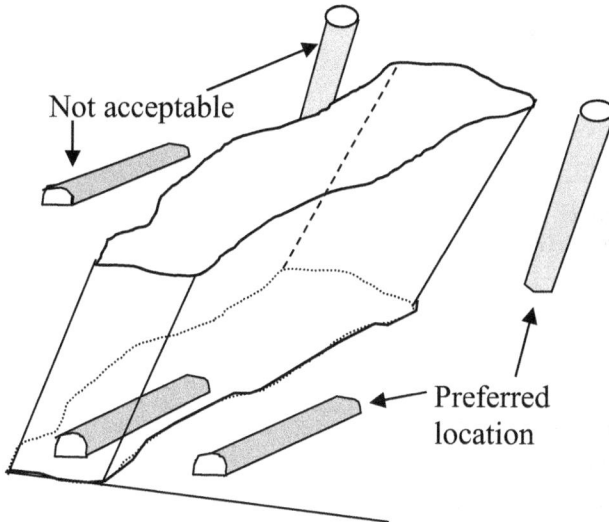

Fig. 8.3 Secondary mine structure locations.

8.2 Optimum Orientation of Mine Openings

Mine openings can be considered in two categories: production openings (e.g., stopes, panels) and mine drifts (including all accesses required to enter production openings). They are discussed separately here.

Mine drifts orientation relative to major in-situ stress direction

Mine drifts may be located in the surrounding rock mass or in the orebody as necessary for production. Those mine drifts, other than the primary and secondary accesses, are called tertiary accesses. They are normally used to serve part of a mine or a local area, such as a stope or production face with a relatively short service life. These drifts are generally developed from the secondary accesses and go directly into the production area. Often there may not be many choices in their direction as other mine structures are already in place.

The primary and secondary accesses however have longer service life and often there are choices in their directions, particularly during the early design stage. As indicated in the previous chapter, a drift should be parallel to the maximum field stress direction. This is to reduce stress concentration induced by excavation around the drift and to avoid ground failure thereof.

During the mine design process, the shape and size of an orebody is often an important factor to consider. For steeply inclined to nearly vertical orebodies, such as metal or base metal orebodies, the horizontal dimension is usually small in comparison to the vertical dimension. The primary and secondary accesses are often relatively short and the choice of direction from the footwall (as discussed earlier) to the orebody is limited. However, for low dip angle orebodies, (e.g., less than 20° such as coal seams), the horizontal dimension is usually much larger than the vertical dimension. In this case, there are more opportunities to choose different directions of these drifts to align with the maximum field stress direction.

Figure 8.4 shows a low dip angle orebody where the maximum field stress is perpendicular to the strike. The main drifts are also preferred to be perpendicular to the strike and are apart from each other to avoid mutual interference.

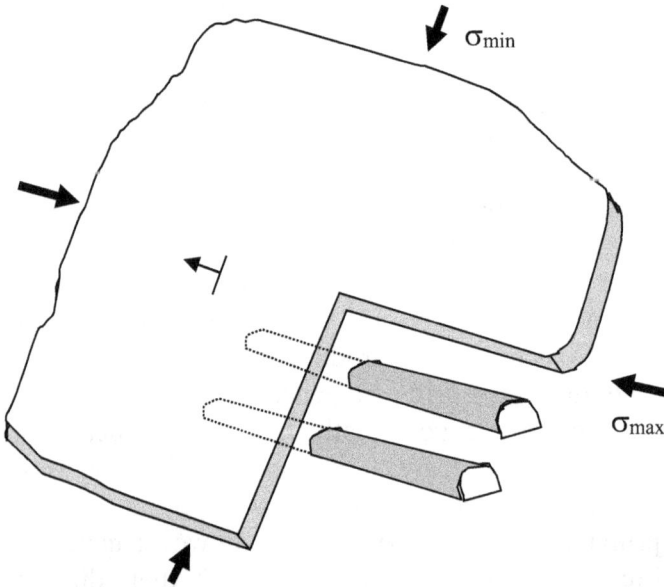

Fig. 8.4 Optimum mine drifts perpendicular to strike.

Figure 8.5 shows a low dip angle orebody where the maximum field stress is parallel to the strike. The preferred main drifts are also parallel to the strike. For this type of orebody, there are more choices to avoid unnecessary high stress concentration induced by excavations.

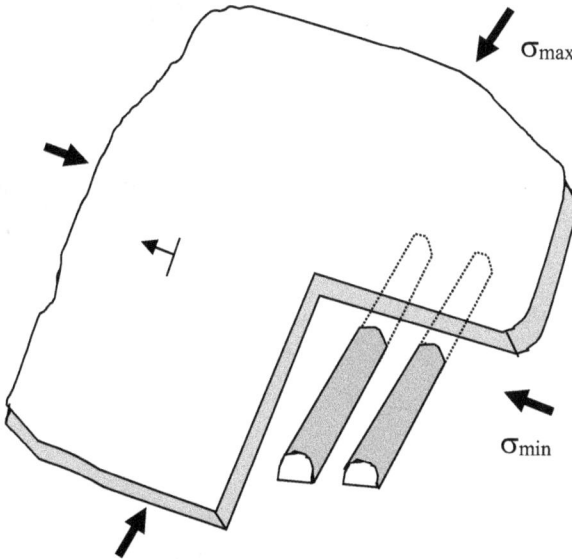

Fig. 8.5 Optimum mine drifts parallel to strike.

Production openings relative to the major in-situ stress direction

The three dimensions of production openings are comparable with each other and the stress re-distribution caused by mine excavation is truly in three dimensions. In this case, they can no longer be simulated by a simplified two-dimensional model and numerical modeling in 3D will be necessary to perform stress analysis in detail.

For steeply inclined to nearly vertical orebodies, the vertical dimension of a stope is usually pre-determined by elevations between two main levels. If the orebodies are narrow, the stope width is usually the same as that of the orebody and the stope length along the strike will be the only flexible dimension. If the orebodies are large with multiple stopes parallel and perpendicular to the strike, both stope width and length will be flexible and can be changed to suit the conditions. Figure 8.6 illustrates preferred pattern of stope layout for two different scenarios of in-situ stresses.

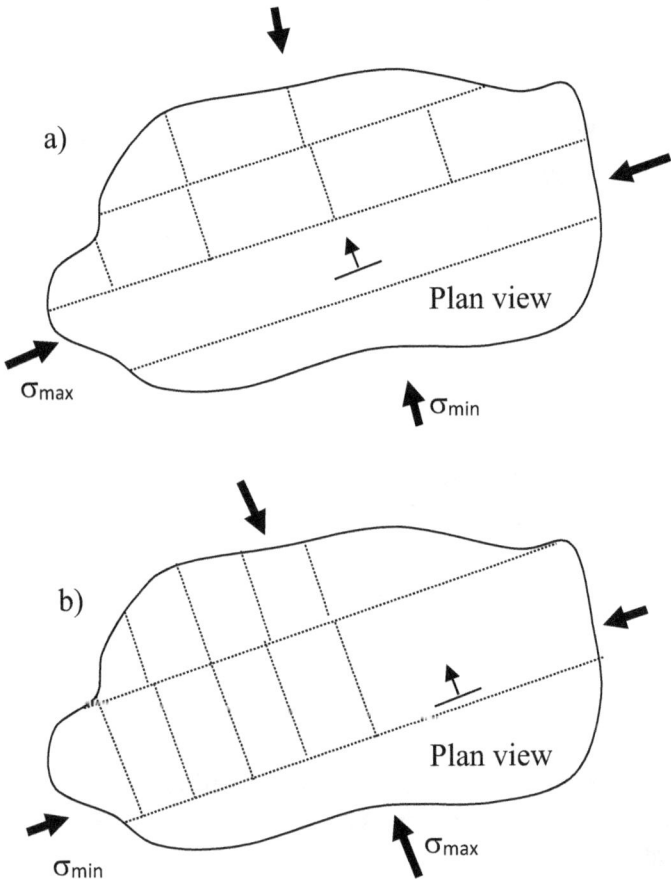

Fig. 8.6 Optimum stope orientation relative to the maximum field stress.

For low dip angle orebodies, (such as a longwall panel or a room-and-pillar panel), the long dimension of the production opening, should normally be in line with the direction of the maximum field stress, (i.e., parallel to the main drifts as illustrated in Figs. 8.4 and 8.5), if other conditions permit.

It is important to point out that the stress condition in a mine is undergoing a continuously changing process as mining progresses. The stope layout demonstrated in Fig. 8.6 is based on the pre-mining stress condition. The field stress state will be different after some stopes are mined out. For example after 1/3 of the orebody is mined, as can be imagined, depending on which 1/3 of the orebody is mined first, the footwall side or the hanging wall side, the stress state on the remaining part of the orebody will not be the same. This is related to mining sequencing. A properly selected mining sequence will help reduce unnecessary stress concentrations significantly. Once again, numerical modelling will help in the process.

Estimate of maximum stress direction

It happens very often that there is little or no information on the in-situ stress state during mine design. It is also a known fact as mentioned before that the original stress field is disturbed and a new stress state is reached on a continuous basis during mining. As a result, the original design may not necessarily be the best option a few years later due to these changes, and modifications will be needed from time to time. It is often not economical to measure the field stresses continuously. Alternative methods are needed to estimate the stress directions.

The direction of the major stress may be estimated from ground failure observations. For example, for the mining stage shown in Fig. 8.7, the vertical blast holes in a stope may exhibit failure sometime later after drilling. Based on the location of the crushing failure, which we know is caused by high compressive stress from discussions in a previous chapter, the direction of the maximum horizontal stress at that stage can be estimated as shown. This may or may not be the original maximum stress direction due to the influence of adjacent mining.

If failure is observed in an ore pass or ventilation raise, which are normally located some distance away from the active production openings, the estimated major stress direction will be closer to that of the original maximum horizontal stress.

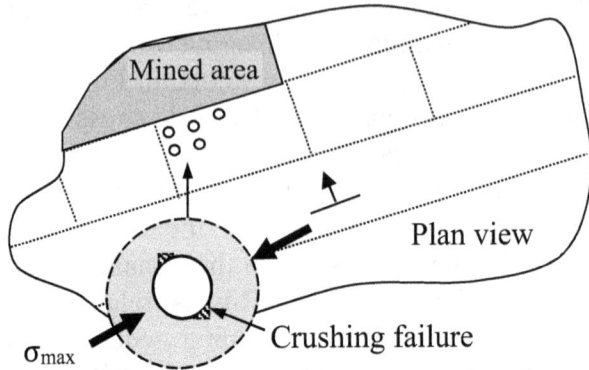

Fig. 8.7 Estimate of major field stress direction from blasting hole deformation.

Although horizontal mine drifts are usually not circular, the above discussed principles still apply. For example, if a horseshoe-shaped drift suffers from high stress related failure at the top left corner (Fig. 8.8), the direction of the maximum stress in the cross section of the drift is most likely slightly off the horizontal line as shown. In this case the horizontal stress may not be a principal stress. However, several factors may contribute to failure at that

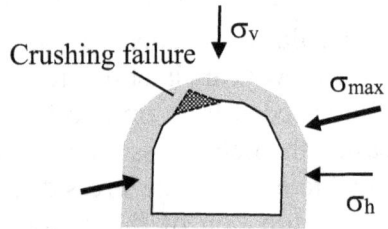

Fig. 8.8 Estimate of the maximum stress direction from drift failure.

location: the in-situ stresses, the shape of the drift (e.g., a corner), local geological condition, and the gravity of the loosened roof rock in relatively weak rock mass. For inclined orebodies, the inclination of the rock may also play a role.

Alternatively, field stresses and directions can also be estimated by back-analysis of field measurements. A 2D method incorporating the boundary element method was developed by Zou and Kaiser (1990). A 3D method based on differential-direction drilling was recently developed by the author and his graduate student Dr. Lin (Lin 2019).

8.3 Open Stope Design with Mathew's Method

The rock mass classification systems, RMR and Q (presented in Chapter 6), have tremendously improved the way of characterizing rock mass quality as well as helped design and support of tunnels and mine drifts. There are however some technical issues when applied to mining production openings. First, mining depth is much deeper than any tunnels and as a result mine structures are under much higher stresses. Secondly, tunnels / mine drifts can be simplified into a two-dimensional problem in terms of stress conditions while mine production openings are in a true three-dimensional stress state. Thirdly, mine production openings have much larger dimensions than tunnels. Therefore, these issues need to be addressed before those classification systems can be adopted for these openings.

A common challenge in mine excavation is to assess the stability of a wall or roof of an open stope. They are most vulnerable around a stope and will cause serious dilution to the ore or completely shut down a stope if they fail pre-maturely. In an effort to assess the stability of the surrounding surfaces of a stope, an empirical relationship between rock mass quality based on the Q-system and stope dimension (span and height) was developed (Golder Associates 1981).

With this method, each surface of a stope (e.g., roof, hanging wall, end walls, etc.) are evaluated. Several geotechnical parameters are used to define a stability number N of a surface. It is then plotted against a shape factor S of the exposed surface. The stability number accounts for the rock mass quality, the active stress and the orientation of the exposed surface. The shape factor accounts for the shape and size of the surface. On this basis, a high stability number and low shape factor reflect more stable conditions while a low stability number and high shape factor reflect less stable conditions.

The stability number N has been applied to numerous field case studies including both failed and stable stopes and the result was a stability chart which can be used to assess other stope stability. The stability of each exposed surface: roof, hanging wall and footwall of a stope, as well as the exposed surfaces of horizontal and vertical pillars, are assessed individually.

In a simple form, the stability number N is defined as

$$N = Q' A B C \tag{8.1}$$

where
- Q' is the modified Q value by setting the stress reduction factor (SRF) to 1.0 with all other factors unchanged in the Q system,
- A is a stress factor,
- B is a joint orientation factor,
- C is an orientation factor of a stope surface.

Stress Factor A:

Factor A accounts for the active stress on an exposed surface and is defined as

$$A = \begin{cases} 0, \text{ if } \sigma_c/\sigma_i < 2, \text{ or if } \sigma_i < 0 \\ 0.1125 \, (\sigma_c/\sigma_i) - 0.125, \text{ if } \sigma_c/\sigma_i \in [2, 10] \\ 1, \text{ if } \sigma_c/\sigma_i > 10 \end{cases} \tag{8.2}$$

where σ_i is the induced stress by excavation within the plane of a stope face (e.g., wall, roof, etc.) under consideration and is calculated at the center of the face (details given below).

When the ratio of intact rock strength to induced stress, σ_c/σ_i, exceeds 10, stress related failure is not expected and any failure under these conditions should be related to movement on geological structures only. In this case, the stress factor (A) is set to a maximum value of 1.0. When the induced stress becomes higher, the strength / stress ratio σ_c/σ_i reduces, increasing the risk of stress related failure. When the strength / stress ratio is reduced below 2.0, the stress factor A is set to zero because by experience, potential instability is expected. A is also set to 0 when tension occurs $\sigma_i < 0$ because of tensile failure.

While the uniaxial strength σ_c can be determined by laboratory tests, the induced stress σ_i on stope surfaces is best estimated by numerical modelling using the in-situ stresses as input data. However, this topic is beyond the scope of this book and readers are encouraged to read relevant publications.

As a simple and quick solution, an approximate approach has been provided and the results are presented in graphic charts in Golder (1981) to help estimate the induced stresses on the stope surfaces. For more detail, users should refer to the original publication. Typical data are presented for the backs, ends and hanging walls of single openings in Table 8.1. Definitions of relevant parameters are illustrated in Fig. 8.9. The process of determining σ_i will be best explained with an example introduced later.

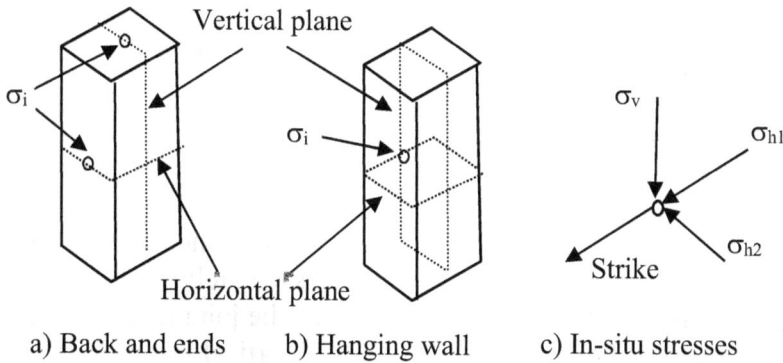

a) Back and ends   b) Hanging wall   c) In-situ stresses

Fig. 8.9 Illustration of induced stresses in back, ends and hanging wall of single stopes.

σ_i – Induced stress,	σ_v – Vertical virgin stress,
σ_{h1} – Horizontal stress // to strike,	σ_{h2} – Horizontal stress $\perp$ to strike

Table 8.1 Estimate of induced stress (After Golder 1981).

a) Ratio of induced stresses σ_i / σ_v on stope back, or σ_i / σ_{h1} on stope ends.

Dimension ratio*	K* = 0.5	1.0	1.5	2.0
1:1	0	0.95	1.90	2.5
2:1	0.45	1.30	2.50	3.4
3:1	0.70	1.80	3.00	4.1
4:1	0.80	1.98	3.25	4.8
5:1	0.90	2.15	3.85	5.4
6:1	0.95	2.30	4.10	5.9
7:1	0.99	2.55	4.45	6.3
8:1	1.0	2.80	4.80	6.6

Table 8.1 (continued)
b) Ratio of induced stresses on hanging wall σ_i/σ_v (vertical plane),
  or σ_i/σ_{h1} (horizontal plane).

Dimension ratio	K = 0.5	1.0	1.5	2.0
1:1	1.26	0.75	0.45	-0.05
2:1	0.90	0.50	0.05	-0.40
3:1	0.75	0.30	-0.15	-0.52
4:1	0.70	0.24	-0.26	-0.60
5:1	0.69	0.25	-0.35	-0.65
6:1	0.65	0.24	-0.38	-0.70
7:1	0.645	0.235	-0.43	-0.74
8:1	0.64	0.23	-0.48	-0.78

* referring to Fig. 8.9, on horizontal planes, $K = \sigma_{h2}/\sigma_{h1}$. On vertical planes, $K = \sigma_{h2}/\sigma_v$. Dimension ratio = long/short of a stope face.

Joint Orientation Factor B:
This factor accounts for the joint effect, determined using Table 8.2, where the roof and wall are treated separately. The value of B depends on the relative angle between the joint plane and the stope surface. For example, for a set of vertical joints, B is 1.0 for the roof and 0.5 for the wall.

Table 8.2 Orientation adjustment factor B (After Golder 1981).

	Joint dip angle $\delta(°)$	B value
Horizontal roof		
	0	0.5
	20	0.3
	45	0.4
	60	0.8
	90	1.0

Vertical walls	Joint dip angle $\delta(°)$	Angle $\beta(°)$ between joint and wall	B value
	90	0	0.5
	70	20	0.3
	45	45	0.4
	30	60	0.8
	0	90	1.0

Surface Orientation Factor C:

Factor C accounts for gravity. It depends on the dip angle, δ, of the surface in consideration. For example, if an orebody dips 70°, the dip angle for the hanging wall is 70° and the dip angle for the horizontal roof is 0°.

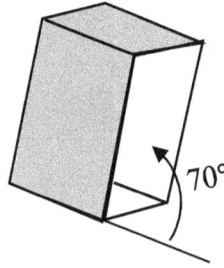

Factor C is calculated by

$$C = 8 - 7 \cos \delta \qquad (8.3)$$

Shape Factor S:

In order to consider geometry and size of a stope face, a shape factor S is introduced, which is defined as

$$S = \text{Area} / \text{Perimeter of the exposed surface} \qquad (8.4)$$

The shape factor S is also called hydraulic radius.

The stability of a stope face is assessed using a stability N-S chart. The chart is plotted on a logarithmic scale for the stability number N ranging from 0.1 to 1000 versus S of an opening surface. The original chart which indicated three zones (stable, transition and unstable) was modified by Potvin (1988) to include more case studies. The two charts are similar but with slight changes in boundaries between zones. Trueman et al. (2000) presented a further modification to include more case studies in Australia.

To use the stability chart, the values of N and S of a stope face are used to find its position in the chart. Depending on the location, an estimate is made whether or not the stope face is stable. As an alternative approximate estimate, typical data based on the work of Potvin (1988) are presented in Table 8.3. Users should however refer to the original publications for more accurate data.

An example is given in the following to demonstrate how to determine the above parameters and how to use this method.

Table 8.3 Stability table (after Potvin 1988).

N value	S<0.2(m)	5 (m)	10 (m)	15 (m)	20 (m)	25 (m)
0.1	s*	pu / u	u ——————————————————→			
1.0	s	s / pu	u ——————————————————→			
3		s / pu	u ——————————————————→			
6		s	pu / u	u ———————————→		
10			s / pu/u	u ———————————→		
20			s / pu	u ———————————→		
40			s	pu / u	u ————→	
70				s / pu	u ————→	
100				s / pu	pu / u	u
200				s	s / pu	u
400					s	pu
700					s	s

* s – stable, pu – potentially unstable, u – unstable.

Example 8.1

An orebody 25 m wide ($\perp$ to strike) dips 85°, with joints dipping 30°. An open stope is to be developed at a depth of 1000 m, cross the whole orebody width, 30 m long and 75 m high. The compressive strength of rock specimens (σ_c) is 120 MPa on average. The estimated in-situ stresses are as follows:

$\sigma_v = 0.027$ (MPa/m) x depth (m),

$\sigma_{ha} = 1.2\sigma_v$, $\sigma_{ha} / \sigma_{hb} = 0.8$,

σ_{ha} // the strike of the orebody.

The rock mass quality Q' is 34.

Assess the stability of the back, hang wall and end of the stope.

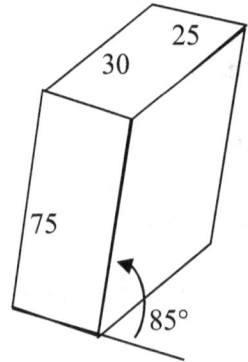

Solutions:

First, draw a stope diagram as shown.

Then, determine the field stresses:

$$\sigma_v = 0.027 \text{ (MPa/m)} \times 1000 \text{ (m)} = 27 \text{ (MPa)}$$

$$\sigma_{h1} = \sigma_{ha} = 1.2\sigma_v = 1.2 \times 27 = 32.4 \text{ (MPa)}$$

$$\sigma_{h2} = \sigma_{hb} = \sigma_{ha} / 0.8 = 32.4/0.8 = 40.5 \text{ (MPa)}.$$

Determine Stress Factor A:

For the stope back, consider the vertical plane in Fig. 8.9a. The field stresses on the plane are σ_v and σ_{h2}.

The dimension ratio on the vertical plane $= 75/25 = 3$,
$K = \sigma_{h2} / \sigma_v = 40.5/27 = 1.5$.

From Table 8.1a, $\sigma_i / \sigma_v = 3.0$,
then $\sigma_i = 3.0 \times 27 = 81$ (MPa)

$\sigma_c / \sigma_i = 120/81 = 1.48$ $\therefore$ A = 0 (Eqn. 8.2)

For the stope end, consider the horizontal plane in Fig. 8.9a. The field stresses on the plane are σ_{h1} and σ_{h2}.

The dimension ratio on the horizontal plane $= 30/25 = 1.2$,
$K = \sigma_{h2} / \sigma_{h1} = 40.5/32.4 = 1.25$.

From Table 8.1a, $\sigma_i / \sigma_{h1} = 1.4$ (estimated),
then $\sigma_i = 1.4 \times 32.4 = 45.4$ (MPa),

$\sigma_c / \sigma_i = 120/45.4 = 2.64$ $\therefore$ A $\approx$ 0.17 (Eqn. 8.2)

For the hanging wall, consider both the horizontal and vertical planes in Fig. 8.9b.

On the vertical plane, the field stresses are σ_v and σ_{h2}. From the above analysis on the stope back,

dimension ratio $= 3$ and $K = 1.5$.

From Table 8.1b, $\sigma_i / \sigma_v = -0.15$,
then $\sigma_i = -0.15 \times 27 = -4.0$ (MPa) < 0 (possible tensile failure noted). However, the value of σ_i is very small, so we need to assess the strength/stress ratio as well.

$\sigma_c / \sigma_i = |120/4| = 30 > 10$, $\therefore$ A = 1.0 (Eqn. 8.2)

On the horizontal plane, the field stresses are σ_{h1} and σ_{h2}. From the above analysis on the end, dimension ratio $= 1.2$ and

K= 1.25.

From Table 8.1b, $\sigma_i / \sigma_{h1} = 0.6$,

then $\sigma_i = 0.6 \times 32.4 = 19.4$ (MPa),

$\sigma_c / \sigma_i = 120/19.4 = 6.2$, $\therefore$ A = 0.57 (Eqn. 8.2)

Use the lower value of A = 0.57 for evaluation of the hanging wall and also keep in mind the small tensile stress.

Determine Joint Orientation Factor B

From Table 8.2, for joints dipping at 30°, we have these values:

    back   0.35 (use average between 0.3 and 0.4)
    hanging wall   0.8 (approximately vertical)
    end   0.8 (vertical wall)

In this example, joints are assumed to have the same effect on the hanging wall and ends, a conservative approach.

Determine Surface Orientation Factor C:

Use Eqn. 8.3, $C = 8 - 7 \cos \delta$
    back $\delta = 0°$, $C = 8 - 7 \cos 0° = 1.0$
    hanging wall $\delta = 85°$, $C = 8 - 7 \cos 85° = 7.4$
    end $\delta = 90°$, $C = 8 - 7 \cos 90° = 8$

Determine Shape Factor S and Stability Number N

With the dimensions of each surface to consider, we can now use Eqn. 8.4 to calculate the shape factor S and use Eqn. 8.1 to calculate the stability number N. The results are summarized below:

	A	B	C	N	Area	Perimeter	S	Stability
back	0	0.35	1.0	0	30×25	2(30+25)	6.8	unstable
hanging wall	0.57	0.8	7.4	114.7	75×30	2(75+30)	10.7	stable*
end	0.15	0.8	8	32.6	75×25	2(75+25)	9.4	transition

* The hanging wall appears stable in terms of compression. It may however become unstable if tension develops in the vertical direction.

The results of N and S for each surface are compared with Table 8.3 (or the stability chart). The following conclusions can be made:

- the stope back is unstable and is most likely to fail in compression because of high stress,
- the stope end walls fall on the boundary between stable and potentially unstable,
- the hanging wall appears stable in terms of compression. It may however suffer from possible tensile failure.

The main concern will therefore be on the back and tension in the hanging wall. Tensile failure in the hanging wall occurs often and causes dilution to the blasted ore. Some dilution is acceptable in practice. One option to avoid failure is to reduce stope dimensions.

8.4 Pillar Design

Mine openings excavated in the solid rock mass leave rock remnants between them, called pillars, in various sizes and shapes. They will have to bear the in-situ stresses unloaded by the openings. The stresses on the pillars will be significantly higher than before. In general, pillars from underground excavations may be simplified to square pillars, rectangular pillars and rib pillars as shown in Fig. 8.10, plus irregularly shaped random pillars. Pillars can be vertical (e.g., those between drifts and stope walls) and horizontal (e.g., sill pillars).

Average pillar stress

The actual stress distribution in a pillar could be complex, varying with pillar size and shape, and may be different at the central section from that near the pillar ends. Numerical modelling is often a useful tool to determine the actual stress distribution. However, for simplicity and for quick reference in design, stress in a pillar is simplified to consist of two components: the average pillar stress and stress concentration. In the following, vertical pillars will be considered for stress calculations.

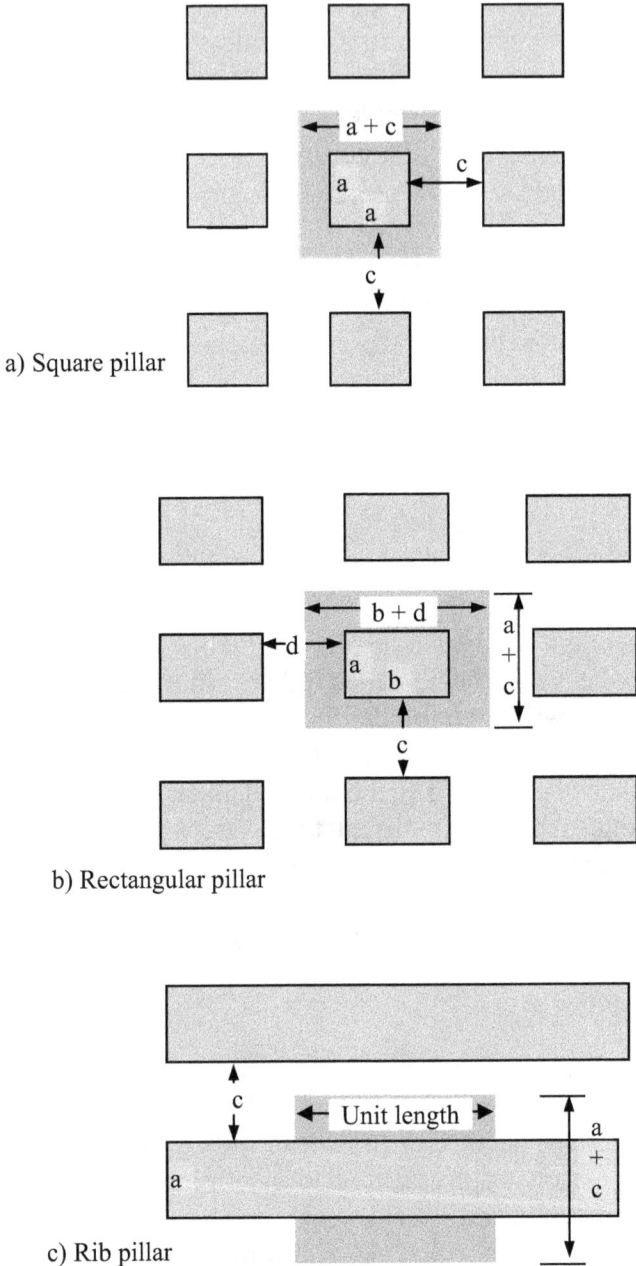

a) Square pillar

b) Rectangular pillar

c) Rib pillar

Fig. 8.10 Typical pillar configuration (plan views).

The average stress in a pillar will depend on pillar size and extraction ratio. The commonly used tributary area theory states that the remaining pillar will have to bear half of the load previously carried by the rock mass in an opening next to it, i.e., half the opened space between two adjacent pillars.

Thus, for a vertical pillar with a definite cross section size A_p, the average vertical pillar stress is determined by

$$\sigma_p = \frac{\gamma z (A_p + A_o)}{A_p} \tag{8.5}$$

where z is the depth below the ground surface and A_o is the effective share of the areas of the openings surrounding the pillar.

For a square pillar with width a and the same distance c to adjacent pillars on all sides (Fig. 8.10a), the average pillar stress is

$$\sigma_p = \frac{\gamma z (a+c)^2}{a^2} = \gamma z (1 + c/a)^2 \tag{8.6}$$

For a rectangular pillar with dimensions a × b and distance c and d to adjacent pillars on both sides (Fig. 8.10b),

$$\sigma_p = \frac{\gamma z (a+c)(b+d)}{ab} = \gamma z (1 + c/a)(1 + d/b) \tag{8.7}$$

For a rib pillar with width a and the distance c to adjacent pillars on both sides (Fig. 8.10c),

$$\sigma_p = \frac{\gamma z (a+c) \times 1}{a \times 1} = \gamma z (1 + c/a) \tag{8.8}$$

For irregular pillars, the pillar area A_p and the effective share of opening area A_o will have to be determined individually based on the actual geometry.

When the above method is applied to shallow excavations and relatively uniform rock masses, reasonable results are expected. However, in deep mine excavations and in rock masses where a major discontinuity exists, the pillar stress may be quite different and will vary with the field conditions.

In deep mine excavation, the load above the excavated area may not be evenly distributed to all pillars due to potential stress arch effect, as demonstrated in Fig. 6.1. In this situation, the pillars under the protection of the stress arch will carry less stress, while the pillars at the abutment of the stress arch will carry higher stress. A typical example is room and pillar mining (Fig. 8.11), where the barrier pillars are expected to carry much higher stress than those pillars in between. In this situation, those pillars between the abutment pillars only need to carry the load under the stress arch, not the entire overburden.

Each situation is different and should be analyzed individually to determine the actual stress in a pillar. Numerical modeling will be a very useful tool to carry out this task.

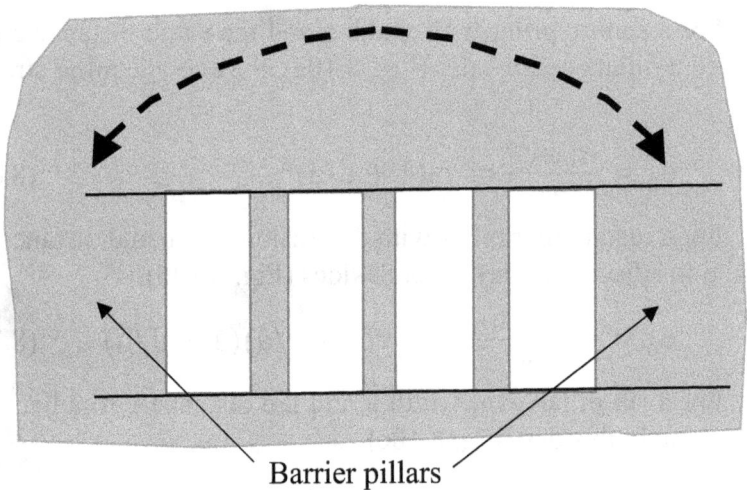

Barrier pillars

Fig. 8.11 Un-even load distribution in Room and Pillar mining

When a major discontinuity exists, the way the load is transferred may be affected. Figure 8.12 illustrates a situation where mining is separated by a fault. The load under the ramp is mostly transferred to the footwall of the fault and the abutment pillar at the bottom of the ramp. The pillars, between the two load carriers, under the ramp may carry very little load and can be much smaller.

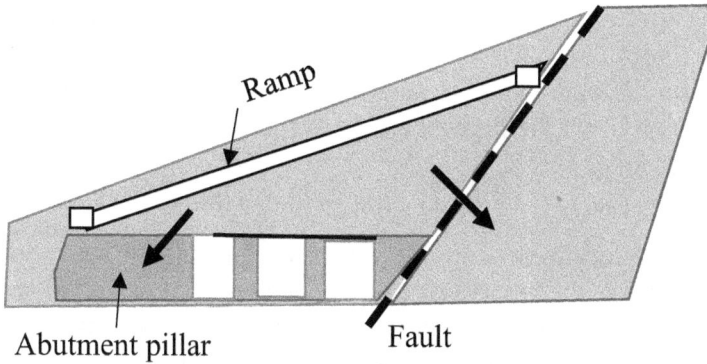

Fig. 8.12 Effect of fault on pillar stress

Pillar stress concentration

The average pillar stress is an approximate estimate of the nominal stress in a pillar. As demonstrated in Fig. 7.10, excavation induced stress redistribution resulted in the highest stress on the opening surface (i.e., the pillar edge) and this concentration affected an area in the rock mass at a distance of 3 to 5 times the opening size. For a pillar this effect exists on all sides where there is an opening, as illustrated in Fig. 8.13. The average vertical stress in the pillar is the same throughout a (horizontal) cross section. The actual stress at a point in the section may be quite different. The stress near the pillar edge could be much higher, and near the center could be lower for a large pillar. The stress in the mid height section may be different as well from that near the ends (top and bottom).

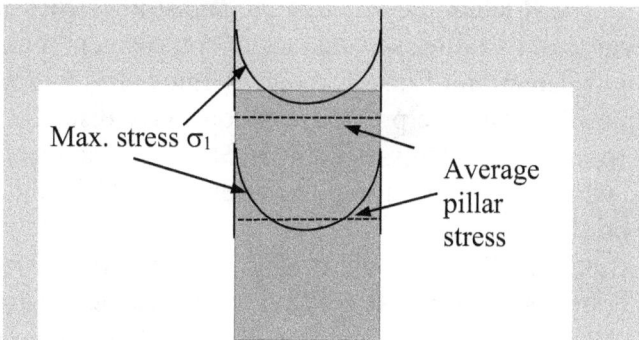

Fig. 8.13 Pillar stress concentration at different locations.

Normally the stress change is compared with the original stress before excavation. The pillar stress concentration factor may be defined as the ratio of the induced vertical pillar stress σ_1 (this will be the highest stress in the pillar) over the average vertical pillar stress,

$$\xi = \sigma_1/\sigma_p, \tag{8.9}$$

This definition can be adopted for sill pillars too by considering the induced and the average horizontal pillar stresses. The stress concentration factor may be less than 1.0 near the center of a large pillar but significantly higher than 1.0 as it gets close to the opening, reaching the highest value on the pillar edge. The smaller the pillar, the higher the stress concentration. For example, at a pillar width / opening span ratio of 5, the concentration factor at the pillar edge could be 6 or more (Hoek and Brown 1980a).

For square and rectangular pillars, the height/width ratio of a pillar has a huge impact on stress distribution in the pillar. Hoek and Brown (1980a) indicated that at a height/width ratio $\geq$ 4:1, stress distribution on a (horizontal) cross section near the middle height of a pillar approaches uniaxial loading condition, i.e., $\sigma_1 \approx \sigma_p$ and $\sigma_3 \approx 0$. As the height/width ratio drops to 1:4, stress becomes non-uniform (both σ_1 and σ_3 vary on the section).

Pillar strength and pillar stability

A rock pillar cannot be treated as a rock specimen because of its large size and potential inclusion of fractures and other discontinuities. It is not the same as the undisturbed rock mass either because of its finite size and lack of confinement rock that masses have. Creation of underground openings has put pillars in a unique situation. The previous sections have discussed the methods for estimating stresses in pillars. From a practical point of view, we need to know the pillar strength and if a pillar is going to be stable or fail.

The strength of a rock pillar will be proportional to the uniaxial compressive strength but at a lower value because of the defects in the pillar and lack of confinement. The empirical failure criterion for a jointed rock mass proposed by Hoek and

Brown (1980b) in Chapter 5 can be used to estimate the pillar strength. The maximum allowable stress, σ'_1, is the strength, which is then compared with the maximum pillar stress (or the average pillar stress × the pillar stress concentration factor).

The pillar is stable, if

$$\sigma_1' / (\xi \, \sigma_p) \geq SF \tag{8.10}$$

where SF is a safety factor. Otherwise, the pillar may fail.

Example 8.2

In a Canadian room and pillar mine, the average square pillar size is 8 m height x 2 m width, with room span of 3 m. At 100 m depth, the rock strength is σ_c = 105 MPa and the estimated GSI = 70. No major discontinuity or water problem exists.
a). Assess the stability of the designed pillar,
b). Is the result applicable to pillars at 1000 m depth in the mine?

Solutions:

a). At 100 m depth, the pillar stress can be estimated with the tributary theory. For a square pillar, by Eqn. 8.6,

$\quad \sigma_p = \gamma H \, (1+c/a)^2 = 0.026 \times 100 \, (1+3/2)^2 = 16.3$ (MPa).

Because H/W = 8 / 2 = 4, the pillar stress in the middle sections is approximately in an uniaxial loading condition, where
$\quad \sigma_1 = \sigma_p = 16.3$ MPa.

The pillar strength can be estimated by the empirical method. By Eqn. 5.15,

$\quad s = \exp((GSI-100)/9) = \exp((70-100)/9) = 0.0357$,

By Eqn. 5.11 for uniaxial compression,

$\quad \sigma_1' \; = \; \sqrt{s} \, \sigma_c \; = \; \sqrt{0.0357} \times 105 = 19.8$ (MPa).

$\quad \sigma'_1 / \sigma_1 = 19.8/16.3 = 1.21$, acceptable for mine pillars.

b). At 1000 m depth, the tributary theory does not apply anymore and the pillar stress should be determined by numerical modelling. The pillar strength σ'_1 and the index GSI should be re-assessed as outlined in Chapter 5 for stability evaluation.

Chapter 9

Support Design for Underground Excavations

Underground excavation disturbs the in-situ stress field and consequently causes stress redistribution and deformation in the rock mass around the excavation. The principal objective of ground support is to help the rock mass support itself by preventing unlimited deformation. Ground support is by no means to stop deformation or to carry the whole load of the rock.

In general, based on the causes, deformation underground can be considered as two types as illustrated in Fig. 9.1:

   a) deformation induced by the loosened rock from geological structures, such as joints and other discontinuities,

   b) deformation induced by stress re-adjustment as a result of underground excavations.

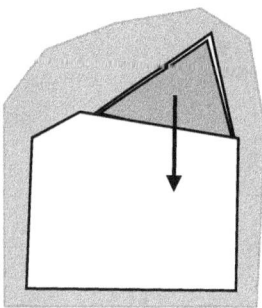

a) Loosened rock

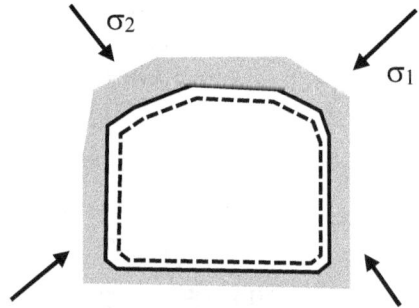

b) Deformation due to stress re-adjustment

Fig. 9.1 Types of ground deformation.

Both types of deformation may exist simultaneously on site. Since the causes of these deformations are different, the required supporting methods will be different accordingly. The first type of deformation will be discussed in the next chapter. The focus below will be on the deformation induced by stress re-adjustment.

9.1 Deformation Caused by Stress Re-adjustment

In a massive rock, what would happen to the surrounding rock mass during and after excavation of an opening? Figure 9.2 illustrates the ground response to excavation and support, and the progress of deformation during excavation. In the example, excavation is completed in sequential steps by the conventional method of drilling and blasting. To simplify this discussion, a hydrostatic stress field with magnitude of σ_o is assumed. Ground supports are installed in tandem sequence after each round of excavation.

Let us look at the cross section X-X:

Step 1. Before excavation reaches that location, the initial stresses (normal and tangent to the boundary line) are in an equilibrium state, $\sigma_{total} = 0$ (Fig. 9.2b, Step 1). When it is far from the excavation face, there is no deformation on the planned boundary line (Fig. 9.2c).

When excavation gets close to it, there may be a small amount of deformation in the radial direction (i.e., perpendicular to the boundary line). This deformation will increase gradually as the excavation face advances closer.

Step 2. When excavation reaches the X-X section, the inner radial stress reduces to zero (Fig. 9.2b, d=0) due to instant removal of rock mass, resulting in stress unbalance $\sigma_{total} > 0$, which would point towards the opening center.

Removal of the inner rock takes place instantly in blasting and an instant large deformation accompanies the excavation in the radial direction (Fig. 9.2c). However, the amount of deformation is restrained by the solid rock ahead of the excavation face.

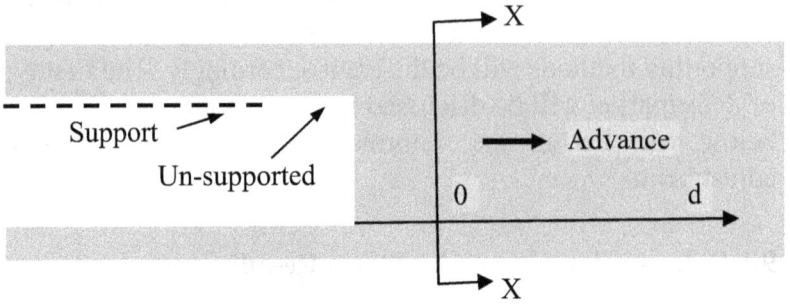

a) Longitudinal sectional view of excavation

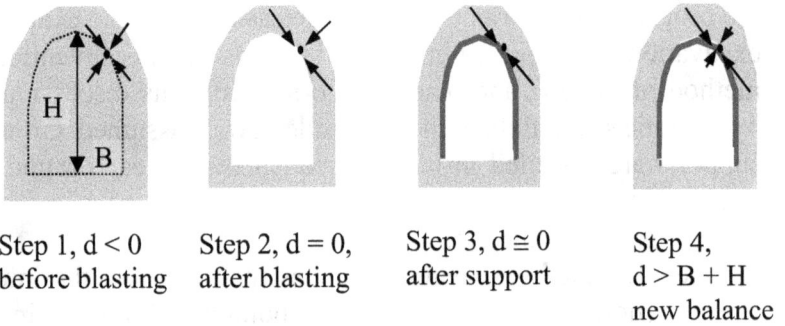

Step 1, d < 0
before blasting

Step 2, d = 0,
after blasting

Step 3, d ≅ 0
after support

Step 4,
d > B + H
new balance

b) X-X cross sectional views in excavation sequence

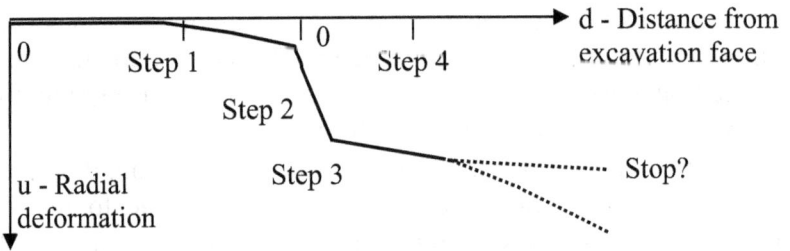

c) Roof movement at section X-X during excavation

Fig. 9.2 Advance of excavation and rock mass responses.
a) advance of excavation, b) stress changes at the boundary of the
opening, before and after excavation, c) roof movement in response to
excavation and support.

As excavation advances passing the X-X section, deformation continues as the stress field readjusts and may stop by itself at some point for strong rock in low stress field.

Step 3. When support is installed if required, deformation may slow down and continue at a low rate as the supporting pressure builds up. Or, it may stop due to the restraint of the support, depending on how much unbalanced stress remains.

From this point, whether the ground will become stable, continue to deform, or collapse will depend on if sufficient support pressure is available and on the rock mass condition.

Step 4. As excavation advances further away, more supporting pressure is built up and more deformation occurs. There are two outcomes from here:

a) If the support eventually builds up sufficient pressure to reach a new stress equilibrium ($\sigma_{total} = 0$), deformation will stop and the rock mass will become self-supportive.

It is also possible that even without support, a strong rock mass in a low stress field may eventually reach stress equilibrium by itself, halting the deformation.

b) If the supporting pressure is insufficient and the rock mass is unable to reach a stress equilibrium state, deformation continues further and the ground fails in various forms.

The concept of deformation and stress balance is further illustrated by a supporting-pressure – deformation curve, shown in Fig. 9.3. The required supporting pressure for the roof (ABDF) is usually higher than that for the walls (ABCG) because of the gravity of a small zone of potentially fractured rock above the roof. Therefore, when the walls become stable, the roof may not even same support pressure is available.

In designing efficient supports, the load-deformation characteristics for both the rock mass and the supporting system must be considered. The rock mass and support system interact with each other and ultimately reach a balanced condition. To reach a balance, the support must build up sufficient pressure in time to meet the demand required by the rock mass.

In the following, the terms pressure and load or force may be used interexchangeably to simplify discussion, although one should keep in mind that pressure is the load per unit area.

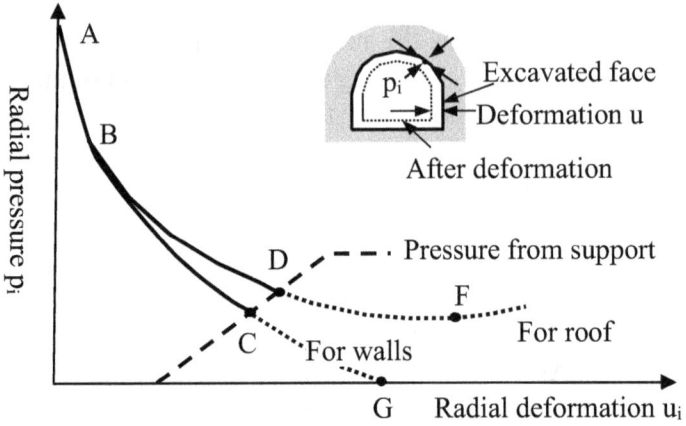

Fig. 9.3 Pressure-deformation curves of roof and walls (Based on Hoek and Brown 1980a)

9.2 Ground Support and Rock Mass Interaction

Assumptions

To demonstrate the concept of how the rock mass and support system interact with each other, the following assumptions are made to simplify the discussion, in reference to Fig. 9.4:

- the opening is circular, with radius r_o,
- the field stress is hydro-static, with $\sigma_1 = \sigma_2 = p_o$,
- the support pressure p_i is uniform on the opening surface,

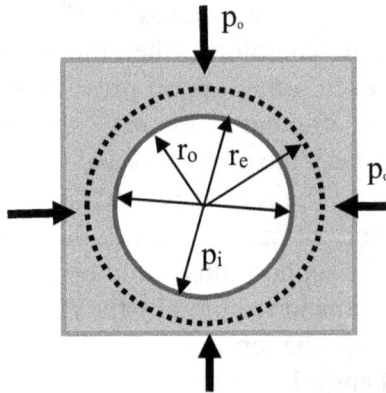

Fig. 9.4 Simplified opening deformation model.

- the rock mass is linearly elastic before failure,
- the rock mass is elasto-plastic with lower strength after fracturing, with a zone size defined by $r_e - r_o$.

The strength properties of the original linear elastic rock and the fractured rock are given by Eqn. 5.10 (the effective stress designation is dropped for simplicity), as

$$\sigma_1 = \sigma_3 + (m_b\, \sigma_c\, \sigma_3 + s\, \sigma_c^2)^a \tag{9.1}$$

where a, m_b and s are the empirical constants for the rock mass. Their values vary with the rock mass quality as described before. The existence of a fractured zone, which extends to radius r_e, will depend on the in-situ stress p_o, the supporting pressure p_i and the rock mass characteristics.

1) The Required support pressure by the rock mass

From theoretical analysis based on the above assumptions, it is shown that a fractured zone (i.e., a plastic zone in this context) may develop when certain conditions are satisfied. The detail is provided by Hoek and Brown (1980a) and only the final results are presented below.

For each type of rock mass, there is a critical pressure, p_{cr},

$$p_{cr} = p_o - M\, \sigma_c \tag{9.2}$$

$$\text{where } M = \tfrac{1}{2} \sqrt{m_b^2/16 + m_b\, p_o/\sigma_c + s} - m_b/8 \tag{9.2a}$$

A fracture zone with radius r_e develops if there is insufficient supporting pressure, i.e.,

$$p_i < p_{cr} \tag{9.3}$$

In this case, the relationship between the radial deformation of the opening, u_i, and the supporting pressure, p_i, is given by

$$u_i = r_o\, [1 - f_1(p_i)] \tag{9.4}$$

where $f_1(p_i)$ is a function of p_i, σ_c, m, s and r_e.

A $p_i \sim u_i$ plot of Eqn. 9.4, as shown in Fig. 9.5, represents "the required support pressure line" for the rock mass.

If sufficient support pressure is available, i.e.,

$$p_i \geq p_{cr} \qquad (9.5)$$

no fracture zone exists (i.e., $r_e = r_o$) and the required support pressure line for the rock mass is given by:

$$u_i = r_o (1 + \upsilon) (p_o - p_i)/E \qquad (9.6)$$

In consideration of the weight of the potential broken rock above the roof, an equivalent pressure w_b is introduced:

$$w_b = \gamma_r (r_e - r_o) \qquad (9.7)$$

where γ_r is the specific gravity of the fractured rock. The sidewalls, roof and floor should then be treated separately. The pressure w_b should be added to (or subtracted from) the p_i value of the required support line for the roof (or the floor). The difference is shown in Fig. 9.5a.

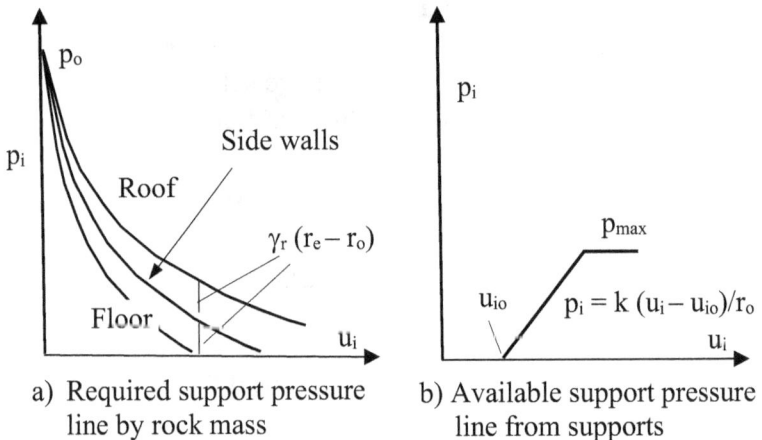

a) Required support pressure   b) Available support pressure
line by rock mass   line from supports

Fig. 9.5 Illustration of required and available support pressure lines (After Hoek and Brown 1980a).

2) The Available supporting pressure from supports

In order to stabilize the ground, the support pressure available from a supporting system must match that required by the rock mass. It should be noted that there is an initial deformation, u_{io}, prior to the installation of any support as demonstrated in Figs. 9.2 and 9.5.

The way a support system is going to respond to an external load is very complex and varies widely underground, depending on the type of support. For example, for shotcrete support, the load from the rock mass may be evenly distributed on the support, while for rock bolts (the commonly used support systems), the load from the rock mass is transformed to tension in the bolt. This transformation is again very complex. The detail of how each type of ground support systems works will not be discussed in this textbook. A simplified model - a spring model, will be used to demonstrate the role of support. If a support system is assumed to work in elastic mode prior to failure, the support line for a support system is given by

$$u_i = u_{io} + p_i \, r_o \, /k \tag{9.8}$$

where k is the stiffness of the support system and it varies with the type of support. Determination of the stiffness of a support is not a simple task. Further information is presented in Hoek and Brown (1980a).

For a specific type of support, there is a maximum allowable pressure or load, defined by p_{max}, which is the capacity of the supporting system. This capacity varies with the type of support. For shotcrete, it may be given as pressure and for rock bolts, it may be given as tension force.

A $p_i \sim u_i$ plot of Eqn. 9.8 represents the available support pressure line from a supporting system, as shown in Fig. 9.5b. Once the support is installed in the ground, pressure builds up in the supporting system as the rock mass continues to deform. When the pressure in the supporting system matches that required from the rock mass, an equilibrium condition is reached and the ground becomes stable.

Ground stabilization requires a proper match of the available and the required support pressures. This is achieved only when the available support line of Fig. 9.5b intersects the required support line of Fig. 9.5a at a proper "time". This timing is represented by the deformation (or location) on the pressure lines. In other words, the required and available pressure lines must intersect each other at the "right location". This "right location" is given in a "time window" (or a range of allowable deformation). They must not intersect too early (the required pressure is larger than the support capacity) or too late (the deformation is already out of control), i.e., at a location on the left of Point F in Fig. 9.3.

The time it needs for the support to build up pressure (or the stiffness of the available support line) after installation must be taken into consideration. If installation is too late, the ground may have failed before the supporting system develops enough supporting pressure. On the other hand, if the support is installed too early, there are two possibilities: a) the available support lines will never intersect the required support line because of inadequate support capacity p_{max}. In this case, the ground may continue to deform as if no support existed; b) the two support lines intersect each other but at a higher pressure point. In this case, the supporting system may function near its full capacity. If the ground deformation continues, the support eventually collapses because of overloading, a result of inappropriate timing, installation being "too early". If the ground deformation slows down and the supporting system can continue to endure further deformation without failure, the timing of installation is "right".

On a mine site, there is more than static loading as discussed above. Mine excavation is a continuous process and so is the ground deformation. Mining often involves blasting, which means instant loosening and removal of a large volume of rock that originally supported the adjacent rock mass. This action creates dynamic loading on the remaining rock mass. In the vicinity of such action, the already balanced ground may become unbalanced again instantly and the supporting system in place will suffer from further (often larger) deformation and may fail.

In hard rock mines, rockburst, the most dangerous type of sudden rock failure involving a large volume of rock, may occur in the following formats. a) Sudden rock mass failure in highly stressed ground (called strain burst), releasing a huge amount of strain energy, and b) sudden slip along a weak geological structure (e.g., fault) caused by reduced normal stress and/or increased shear stress from mine excavations. In the vicinity of rockbursts, the ground will suffer from the highest dynamic loading and the largest deformation in a mine. Again, in such situations, the already balanced ground may become unstable and the supporting system, if not capable of adjusting itself to tolerate larger deformation, may completely fail because of the high pressure.

A rigid supporting system (such as shotcrete) is supposed to provide adequate supporting pressure to the ground and "hold still" to maintain ground stability. This type of support is ideal for low stress and static loading conditions, in areas far from active mine excavation and without further deformation. In areas near the active mine excavation, such as cross cuts to a stope, or in a burst-prone mine, this type of support is not suitable and should be avoided due to its inability to adjust to changing conditions. An ideal support in this situation should be able to yield to large deformation without losing its supporting capacity.

As can be seen, in selection and installation of ground support, all three of these factors will play a role: proper type of support, support capacity and timing of installation. Any of them may cause failure. That is why a type of support working effectively in one mine may not work in another mine. In order to provide effective ground support, it is important to understand the interaction between the rock mass and the support. For convenience, some typical support parameters provided by Hoek and Brown (1980a) are shown in Table 9.1. More detail is available in their original publications.

3) An example of deep mine application

In the following section, a case study given by Hoek and Brown (1980a) is used to illustrate applications of rock-support analysis in deep mines.

Consider a circular mine drift of 4 m radius in Quartzite of very good quality at 1000 m below the ground surface. The rock mass properties are, $\sigma_c = 300$ MPa, $E = 40$ GPa and $\upsilon = 0.2$. To simplify the issue, a hydrostatic stress condition is assumed on the site (i.e., $\sigma_v = \sigma_H = \sigma_o$). Since the stress is high in deep

Table 9.1 Typical support capacity for various support systems (after Hoek and Brown 1980a).

Supports \ Tunnel radius	r_o 1 m 39"	r_o 2.5m 98"	r_o 5 m 197"	r_o 10m 394"
A. Shotcrete - 5cm / 2" thick, σ_c = 14 MPa / 2000 psi after 1 day.	p_{max} 0.65 MPa 95 psi	p_{max} 0.27 MPa 39 psi	p_{max} 0.14 MP 20 psi	p_{max} 0.07 MPa 10 psi
B. Shotcrete - 5cm / 2" thick, σ_c = 35 MPa / 5000 psi after 28 days	1.63 MPa 236 psi	0.68 MPa 99 psi	0.34 MPa 50 psi	0.17 MPa 25 psi
C. Concrete - 30cm / 12" thick, σ_c = 35 MPa / 5000 psi after 28 days	7.14 MPa 1036 psi	3.55 MPa 515 psi	1.93 MPa 279 psi	1.00 MPa 146 psi
D. Concrete - 50cm / 19.5" thick, σ_c = 35 MPa / 5000 psi after 28 days	9.72 MPa 1410 psi	5.35 MPa 775 psi	3.04 MPa 440 psi	1.63 MPa 236 psi
E. Light steel sets - (6 I 12) space 2m / 79", blocked, σ_{ys} = 248MPa/36,000 psi	0.61 MPa 88 psi	0.18 MPa 27 psi	0.07 MPa 10 psi	0.02 MPa 3 psi
F. Medium steel sets - (8 I 23) space 1.5m / 59", blocked, σ_{ys} = 248MPa / 36,000 psi	1.59 MPa 230 psi	0.50 MPa 72 psi	0.18 MPa 27 psi	0.06 MPa 9 psi
G. Heavy steel sets - (12 W 65) space 1m / 39", blocked, σ_{ys} = 248MPa / 36,000 psi	7.28 MPa 1055 ps1	2.53 MPa 366 Psi	1.04 MPa 150 psi	0.38 MPa 55 psi
H. Very light rock bolts – 16mm / 5/8" Ø, at 2.5m/98" centres. Mechanical anchor. T_{bf} = 0.11MN /25,000 lb.	0.02 MPa 2.6 psi	0.02 MPa 2.6 psi	0.02 MPa 2.6 psi	0.02 MPa 2.6 psi
I. Light rock bolts – 19mm / ¾" Ø, at 2m/79" centres. Mechanical anchor. T_{bf} = 0.18MN /40,000 lb.	0.045MPa 6.5 psi	0.045MPa 6.5 psi	0.045MP 6.5 psi	0.045MP 6.5 psi
J. Medium rock bolts – 25mm / 1" Ø, at 1.5m/59" centres. Mechanical anchor. T_{bf} = 0.267MN /60,000 lb.	0.12 MPa 17 psi	0.12 MPa 17 psi	0.12 MPa 17 psi	0.12 MPa 17 psi
K. Heavy rock bolts – 34mm / 1 3/8" Ø, at 1m/39" centres. Resin anchored. T_{bf} = 345MN /150,000 lb.	0.34 MPa 49 psi	0.34 MPa 49 psi	0.34 MPa 49 psi	0.34 MPa 49 psi

ground, the weight of the "dead rock" above the roof is ignored because of its small magnitude. Due to the proximity to the production area, it is required to investigate the stability of the drift and the support measures under different stress scenarios.

Solutions:

Opening radius r_o = 4 m. The average vertical stress at 100 m depth is

$\sigma_v = \gamma z = 0.027$ (MPa/m) $\times$ 1000 (m) = 27 MPa.

Four types of stress scenarios are considered:
A, $p_o = \sigma_v = 27$ MPa
B, $p_o = 2\sigma_v = 54$ MPa
C, $p_o = 3\sigma_v = 81$ MPa
D, $p_o = 4\sigma_v = 108$ MPa.

The rock mass constants are then determined following the procedure given in Section 5.5:
For the original rock, m = 7.5, s = 0.1, and
for the fractured rock, m = 0.3, s = 0.001.

The next step is to determine the critical pressure p_{cr} using Eqn. 9.2 and the $p_i \sim u_i$ relationships using Eqns. 9.4 and 9.6 for each of the four scenarios. Dimensionless support lines $p_i/p_o \sim u_i/r_o$ are plotted for the rock mass. An example is shown in Fig. 9.6. Each support line has two segments joining at $p_i/p_o = p_{cr}/p_o$, represented by the dashed line.

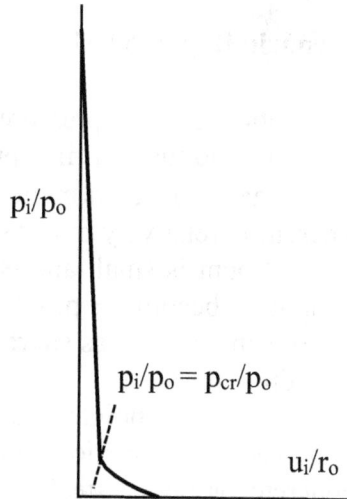

Fig. 9.6 Sample analysis results (after Hoek and Brown 1980a).

Critical data of the analysis results by Hoek and Brown (1980a) are listed in Table 9.2. Based on the results, the stability of the drift can now be assessed and the support requirement can be determined for each type of stress condition.

Table 9.2 Critical data for the four scenarios in deep mine application (After Hoek and Brown 1980a).

Scenario	p_0 (MPa)	p_{cr} (MPa)	Max u_i
A	27	0	(elastic range)
B	54	1.0	0.22% (8.8 mm)
C	81	6.0	0.98% (39 mm)
D	104	13	4.4% (176 mm)

Scenario A: $p_0 = 27$ MPa

The deformation is in elastic range and the critical support pressure $p_{cr} \approx 0$. No fracture zone exists and no support is required.

Scenario B: $p_0 = 54$ MPa

The critical support pressure $p_{cr} \approx 1.0$ MPa. Some fracture will occur around the drift if supporting pressure is below 1.0 MPa. The maximum deformation is $u_i \approx 0.22\%$ $r_0 = 8.8$ mm. Therefore, relatively minor spalling may occur without support. This amount is small and is considered tolerable in a mine. If rock falls become a problem due to fracturing and spalling, minor support such as short bolts with screens will be adequate to address the issue.

It should be pointed out that it is not economic to prevent non-elastic deformation. As shown in Table 9.1, 30 cm thick concrete or heavy steel sets is required to provide 1.0 MPa support pressure. That will be costly.

Scenario C: $p_o = 81$ MPa

The critical support pressure $p_{cr} \approx 6$ MPa, and the maximum deformation $u_i \approx 0.98\%$ $r_o = 39$ mm without support. Severe spalling and fracturing are expected and support is required. It is however nearly impossible to provide 6 MPa of support pressure with any type of support as shown in Table 9.1.

Concrete lining or heavy steel sets, which would supply a support pressure of 1 to 2 MPa, appear to be an obvious solution to this problem. However, they are not economical for this application. In this case, the proper type of support to be installed is to control spalling rather than stop deformation since the deformation is eventually going to stop and the drift will reach a state of equilibrium without support.

If the fractured rock can be kept in place to prevent progressive ravelling, there is little danger of the fracture zone propagating to the point of collapse. A proper solution to this problem is to use light support, to be installed close to the excavation face to prevent progressive ravelling and spalling as well as make the rock mass self-supportive. Screens may be used if small pieces of rock are a problem. Light steel sets with relatively soft blocking would also be acceptable. Use of un-tensioned grouted rock bolts or "split sets" is even more economic. More discussion will be given later on the behaviour of supporting systems.

There have been examples where attempts to control large deformation by use of heavy support were unsuccessful and resulted in support failure. This was avoided in a similar situation by use of un-tensioned grouted rock bolts as a solution. This type of support allowed deformation and therefore relaxation in the ground before taking on a load.

Scenario D: $p_o = 104$ MPa

The critical support pressure $p_{cr} \approx 13$ MPa, and the maximum deformation $u_i \approx 4.4\%$ $r_o = 176$ mm. A severe problem is expected in this case. Heavy supports are needed. Grouted reinforcing bolts plus wire mesh are recommended to control spalling. An alternative or a better option is to reduce the opening size if operation requirements allow.

4) An example of shallow excavation application

Another example provided by Hoek and Brown (1980a) is a circular tunnel of 5.33 m radius designed in a fair quality gneiss at 122 m below the ground surface. The vertical stress in the field is estimated at 3.3 MPa. It is required to design adequate support for the tunnel.

Solutions:

In this case, the weight of the loosened rock above the roof is important and should be considered because the total stress in the ground is not very high. Follow the same procedure as in the previous example to determine the required support line for the sidewalls. Add to (or subtract from) this line a pressure equal to the weight of the loosened rock to determine the required support line of the roof (or the floor). The pressure – deformation lines for the roof and wall are illustrated in Fig. 9.7.

Based on calculation, the critical support pressure $p_{cr} \approx 0.67$ MPa. The support curves for both the roof and wall are at pressure < 0.6 MPa, indicating existence of a fractured (non-elastic) zone. The sidewall will develop more than 100 mm of deformation without support but eventually stops. However, the roof deformation will never stop if supporting pressure is less than 0.09 MPa or not provided before 75 mm of deformation. Supports are therefore needed for the wall and roof.

If a minimum of 25 mm deformation is tolerable, the maximum required support pressure is ≤ 0.22 MPa for both roof and sidewall. We can accordingly select the appropriate supports from Table 9.1. Several options of support are available: steel sets, shotcrete and rock bolts. However, it is clear that the performance of each type of support also depends on the timing of installation. Selection of a proper support and proper timing in installation are equally important.

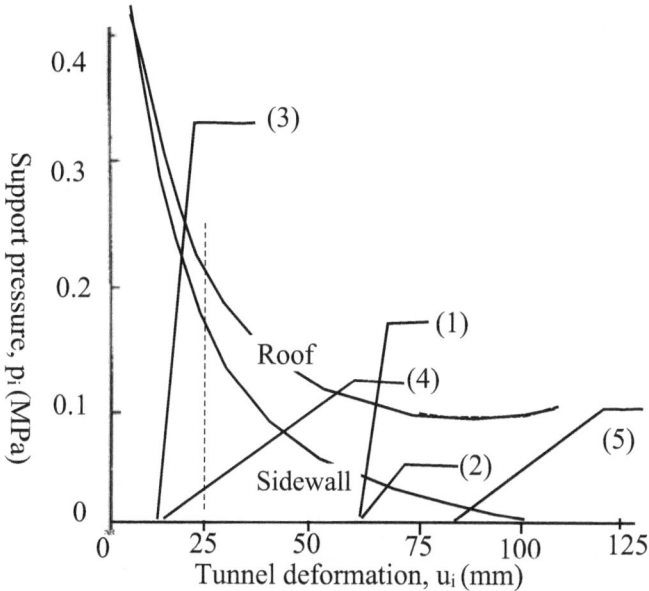

Fig. 9.7 Various supporting strategies (after Hoek and Brown 1980a).

Notes: (1) 8I23 Steel sets at 1.5m centers with good blocking
(2) 8I23 Steel sets at 1.5m centers with poor blocking
(3) 50mm thick shotcrete
(4) 25mm diameter 3m long mechanically anchored rock bolts at 1.5m spacing within 3m of face
(5) 25mm diameter 3m long mechanically anchored rock bolts at 1.5m spacing about 9m from face

a). Conventional choice: Use of medium steel sets (8I23, F in Table 9.1) at 1.5 m spacing, installed at 60 mm initial deformation because installation of steel sets needs more space to work with than rock bolts and shotcrete.

The supporting capacity for an 8I23 steel set p_{max} = 0.18 MPa seems adequate after 60 mm deformation as shown in Fig. 9.7. However, use of steel sets needs blocking (as shown in the margin) between the steel sets and the rock.

The quality of blocking will also affect the performance of the supporting system. In Fig. 9.7, the effect of good blocking – line (1) and poor blocking - line (2), are demonstrated. As shown, for the exact same support system installed the same way, poor blocking renders the support ineffective.

Note: 1.5 m spacing may be too large in blocky ground. Decreasing spacing may help but is costly! Wire mesh or shotcrete may be used as secondary support to hold small rocks.

b). Alternative 1: Use of 25 mm thick shotcrete (B in Table 9.1). The supporting capacity p_{max} is 0.34 MPa, which is adequate, a surprisingly good result, as shown as line (3) in Fig. 9.7.

Because shotcrete can be applied earlier than steel sets after excavation, the initial deformation prior to installation of support is smaller, say, at 18 mm deformation. It provides a practical and economical alternative to steel sets, surprisingly with much better performance.

Note: shotcrete is brittle and rigid. It is therefore:

- not good for places where stress changes constantly, such as draw points, or burst-prone ground. Use of wire mesh or adding steel/glass fibre to a shotcrete mix will help reinforce the shotcrete but will increase the cost.

- not recommended if the excavation profile is irregular because uniform stress distribution on the lining cannot be assured.

c). Alternative 2: Use of medium rock bolts, 25 mm diameter, mechanically anchored at 1.5 m spacing (J in Table 9.1). The supporting capacity p_{max} is 0.12 MPa, also adequate as shown by the support line (4) in Fig. 9.7.

Rock bolts can be installed earlier than steel sets. However, installation of rock bolts requires consideration of other factors:

- timing: the stiffness of a whole rock bolt system is lower than shotcrete. Even if it is installed at the same time as shotcrete, it will take a longer time for rock bolt to develop adequate support pressure, if not pre-tensioned. The supporting capacity is not fully mobilized until the bolt is tensioned by further ground deformation. Therefore, rock bolts should be installed as early as possible. Late installation renders the support ineffective as shown by line (5) in Fig. 9.7.

- pre-tension: if immediate support is required to support the ground, tension needs to be added to the bolt at installation.

- roof and walls: because of the potential broken rock above the roof, more support pressure is required for the roof than the walls. Reduced spacing in the roof may be a solution.

- bolting pattern: if various support pressures are required at different locations, the bolting pattern (e.g., spacing between bolts) can be changed to achieve the objective.

9.3 Empirical Approaches to Underground Support Design

The analytical approach discussed above on ground support helps us understand the mechanism and interaction between the rock mass and the supporting system. It may, however, not be easy to implement without the required knowledge and information. An alternative for underground support design is the empirical methods by using ratings from rock mass classification systems, such as the RMR and Q.

1) Estimate of the maximum unsupported span, S

In a specific ground condition, there is a maximum allowed span, S, without support. If the excavation span is greater than S, ground support will be required.

Barton (1976) compiled a number of cases of man-made and natural unsupported excavation in rock masses with different qualities. The results indicate that no support seems to be required if the span satisfies

$$S \leq 2\,Q^{0.66} \qquad\qquad (9.9)$$

Based on the results, Barton extended the findings to define the maximum unsupported spans for different types of openings, from temporary mine openings to long term underground storage facilities. An excavation support ratio (ESR), similar to a safety factor, was introduced for different applications (Table 6.14).

202 D.H. Steve Zou

Equation 9.9 can be written as a log function

Log S = Log2 + 0.66LogQ = 0.301+0.66LogQ

Taking ESR into consideration, we have a general formula

Log S = a + b Log Q   (9.10)

where a and b are constants, varying with category A to E based on ESR values. Equation 9.10 represents a linear relationship between log S and Log Q. For applications in categories A to E, this relationship is shown in Fig. 9.8. The constants a and b are given in Table 9.3.

There are other similar empirical methods to determine the unsupported span in various applications. Some examples are given in the literature (e.g., Lauffer 1958, Merritt 1972, Bieniawski 1989, NGI 2013).

Table 9.3 Constants of Eqn. 9.9 for different applications

Category	A	B	C	D	E
ESR	3~5	1.6	1.3	1.0	0.8
a:	9.43	3.85	2.97	2.38	1.75
b:	0.39 (same value for all cases)				

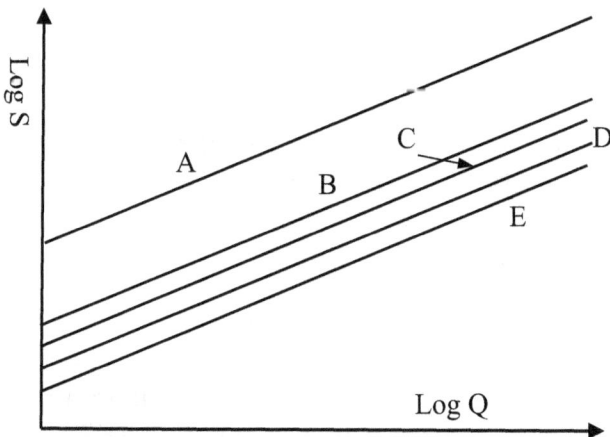

Fig. 9.8 Relationship between Q and unsupported opening span S
(Based on data in Barton 1976).

2) Use of rock mass classification systems for support selection

One of the primary applications of rock mass classification systems is to help users estimate the requirement of support and select proper ground support systems based on rock mass quality in various applications.

In the NGI system, the Q value is used to estimate support requirement as shown in Table 6.15. There is also a very detailed guideline in Hoek and Brown (1980a, Tables 18 and 19) to help select specific supports and support patterns.

It should be noted that those recommendations are a starting point in selecting mine supports. They were mainly used for tunnels at relatively shallow depth, mostly ≤ 1000 m. The database may not cover all cases in North America. Users should exercise caution when applying to deep underground mine excavations.

9.4 Commonly Used Underground Support Systems

Types of available ground support systems

There are a variety of ground support systems available on the market at present. The commonly used supports will be introduced in the following sections.

1) Rock bolts

Rock bolts are the most commonly used supporting system in underground excavations. Figure. 9.9 shows some typical rock bolts installed underground as reinforcement.

A rock bolt used in rock engineering projects is usually made of a steel stem (often a rebar or hollow tube) and anchoring mechanisms, which can be very different depending on the installation method. Based on the make-up and installation method, rock bolts can be divided into these three categories: mechanically anchored, frictionally anchored and grouted (partially or fully). Based on the initial load bearing condition, they can also be grouped as pre-tensioned or un-tensioned, depending on the strategy of ground stability control.

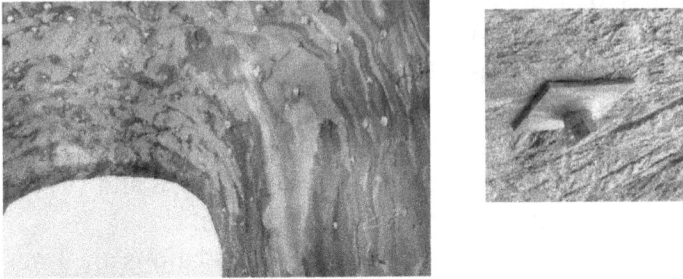

Fig. 9.9 Examples of installed rock bolts.

a) Mechanically anchored rock bolts

This type of rock bolt has several components (Fig. 9.10): a rebar, a nut, a face plate and a bottom anchorage which may be an expansion shell or a wedge. In installation, the rebar is anchored by mechanical means at both ends in a pre-drilled hole in the rock mass and the rest of the rebar is free, with no contact with the rock mass. The bottom anchor secures the rebar to the bottom of the hole. The nut and the face plate make up the anchor at the collar of the hole with the face plate against the rock surface.

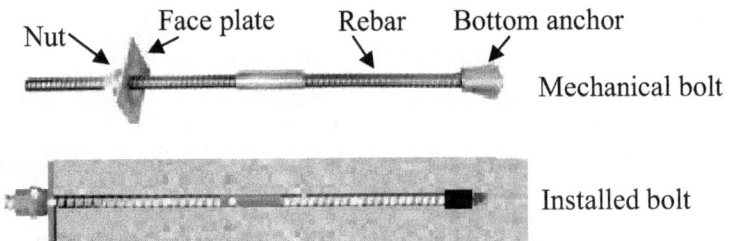

Fig. 9.10 Mechanically anchored rock bolt.

Depending on the needs, the rebar may be pre-tensioned to up to 70% of its capacity by tightening the net. Even without pre-tension, a tensile force will develop in the rebar as the ground moves. The tension increases as ground deformation continues. It will take less time for a pre-tensioned bolt than an un-tensioned bolt to reach full capacity. This force is transmitted as a clamping force to the rock mass between the anchors, thereby restricting ground movement.

For mechanical bolts, the tension is uniformly distributed along the bolt between the anchors. Zou (2004) presented a detailed analysis on tension mobilization and distribution in different types of rock bolts. Tension distribution for mechanical bolts is shown in Fig. 9.11.

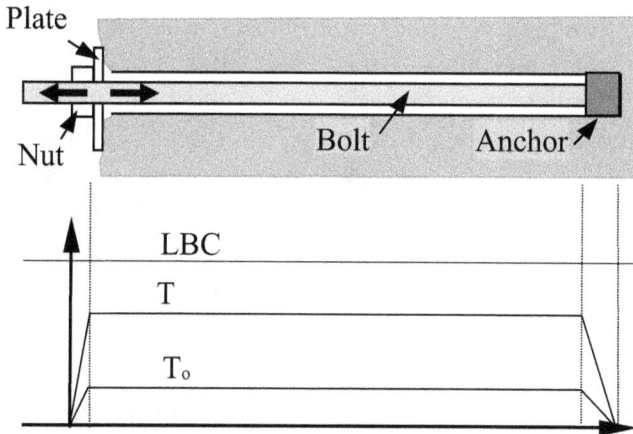

Fig. 9.11 Tension distribution in mechanical bolts (After Zou 2004). LBC stands for the maximum load bearing capacity, T for tension and T_o for initial tension.

The components of a mechanical rock bolt all together make up the supporting system. The supporting capacity of a rock bolt system is only as strong as its weakest component. Therefore, no matter how strong the rebar is, its capacity is limited by its weakest component. Failure of any component will render the whole system ineffective. Component failure may include: nut failure at its thread mechanically or by corrosion, face plate failure by crushing or corrosion, rebar failure by rupture under high tension or corrosion, bottom anchor failure mechanically or by corrosion, and crushing of rock mass at the anchoring location. Failure at the anchoring point is more likely to occur in soft rocks. As a result, use of high strength rebar in soft rocks becomes meaningless. As can be seen, corrosion may be the biggest threat to the free rebar. In some cases, the rebar is coated to work against corrosion. In addition, improper installation of the bottom anchor may also affect the capacity of a rock bolt system.

Mechanical rock bolts are easy to install. Often drilling and installation are completed at the same time with no delay in between. This can provide early support to the freshly exposed ground. It is good for use in medium to hard rocks, but not recommended for soft rock or a corrosive environment.

The face plate comes in different shapes and strengths. Some are flat, some are dome-shaped. The latter is designed to allow the rock bolt system to yield (i.e., to deform slightly) when excessively high stress is developed in the bolt, allowing larger deformation to take place while maintaining its supporting capacity. There are other types of designs of yielding bolts. They are good for burst-prone ground and in places where a large deformation is expected during mine operation.

b) Frictionally anchored rock bolts

This type of bolt is simply a steel tube and a face plate (Fig. 9.12). The steel tube is anchored in the pre-drilled hole by tight contact between the rock mass and the tube. There are two types of friction bolts: split set and Swellex bolt. The split set is a

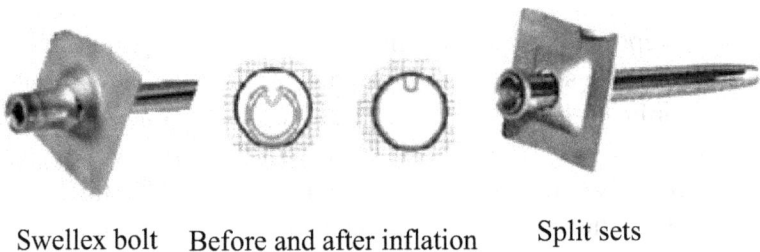

Swellex bolt   Before and after inflation   Split sets

Fig. 9.12 frictionally anchored rock bolts.

slotted steel tube and is slightly larger than the hole. The tube is hammered into the hole by force, creating a tight contact with the rock mass. For the Swellex bolt, the steel tube is deformed in cross section and is smaller than the hole. During installation, the tube is first inserted in the drilled hole and then inflated by high pressure water. The inflated tube will have tight contact with the rock mass and conforms to the shape of the hole.

Frictional rock bolts cannot be pre-tensioned. Supporting force is activated by ground movement. A frictional force is developed at the contact points between the tube and the rock surface of the hole as the rock mass moves. The frictional force restricts ground movement. Since the rock surface of the drilled hole is not very smooth to create a "perfect contact" with the tube, the frictional force is not uniformly distributed along the tube. In reality, this force is most likely the total friction at numerous contact points along the tube.

Since the supporting force depends on the friction, some rust on the tube surface may help increase the supporting capacity. However, excessive corrosion will cause the tube to fail. Failure of friction bolts is usually caused by slip at the contacting points, due to either excessive corrosion of the tube or crushing of rock at the contact points. Large forced ground movement may destroy the rock contacts, rendering the bolt ineffective.

Friction bolts in general have lower supporting capacity than other bolts. They are often used in medium to relatively low stress conditions, and in a non-corrosive environment. They are reported to be good for mild burst-prone ground and large deformation as well.

c) Grouted rock bolts

In this case, a rebar is anchored in the rock mass by grouting (other than mechanical anchors), with or without a face plate and a tightening nut, depending on the application requirement.

Grouting may be along the entire bar or only part of it. Figure 9.13 shows a partially grouted rock bolt with a face plate and illustrates tension distribution along the bolt.

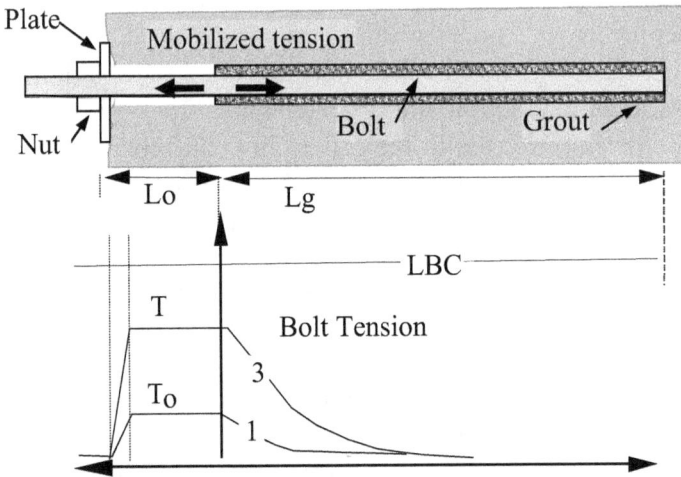

Fig. 9.13 Tension distribution in grouted bolts with bearing plate in a
uniform rock mass (After Zou 2004).

Experiments showed that one foot long grouting will create
approximately sufficient bond to match the tension capacity of a
20 mm diameter steel rebar. Therefore, if the purpose of grouting
is only to create an anchor in the rock mass, there is no need to
fully grout the bolt.

The grouting material may be cement, which takes longer to
cure, or packaged resin, which may cure in less than one hour.
With cement grouting, there are two methods: the breathing tube
method (Fig. 9.14a) and the grouting tube method (Fig. 9.14b).
In the former, grout is pushed in the hole from the collar to the
bottom and air is allowed to bleed through a tube. In the latter,
grout is delivered to the bottom of the hole through a tube, which
is retreated backwards as grout fills the hole.

With packaged resin grouting (Fig. 9.14c), resin packages
are first inserted in the drilled hole and a rebar is pushed in while
spinning to break the resin package and mix it. If pre-tension is
required, the bottom package of resin will set first, in less than
15 minutes. After the bottom resin sets, the nut is tightened
against the plate at the collar to create tension in the bolt. Later
setting of the rest of the resin will securely bind the bolt to the
rock mass.

a) Breathing tube method

b) Grouting tube method

c) Grouting with packaged resin

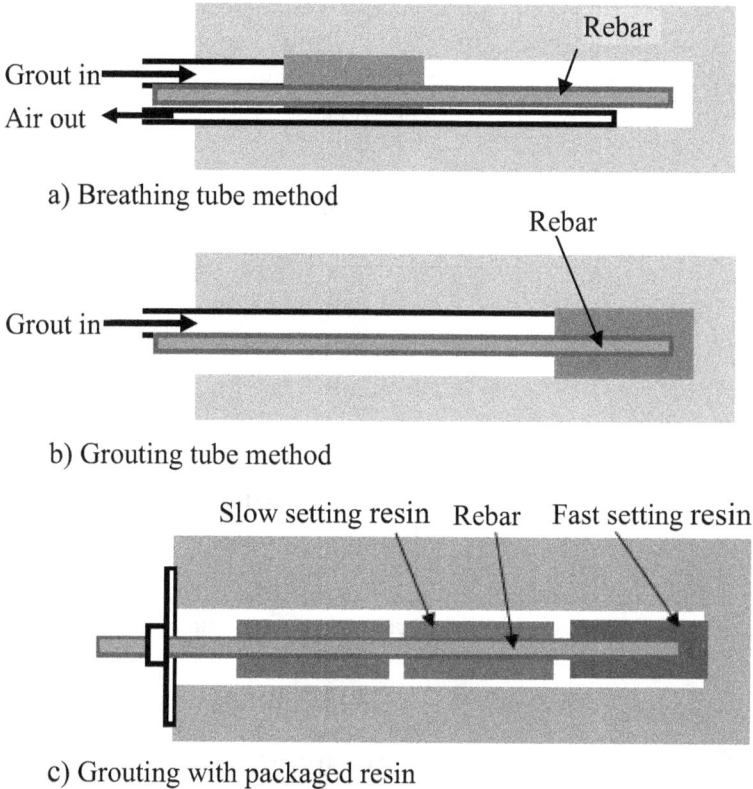

Fig. 9.14 Grouting methods for rock bolts.

For grouted rock bolts, tension mobilized in the bolt is distributed along the bolt in a complex manner (Zou 2004). If a face plate is installed with a partially grouted bolt (Fig. 9.13), the free section of the bolt bears evenly distributed tension. In the grouted section, however, tension is distributed in a decreasing scale. In other words, tension in some sections of the bolt where larger deformation has occurred may be high while there is no tension in the bottom section.

If no face plate is installed, there is no pre-tension in the bolt. Supporting pressure is only activated by ground movement. Tension is not uniformly distributed along the bolt, as shown in Fig. 9.15. The highest tension is at the location where larger movement occurs first, particularly when there is a weak joint intersecting the bolt. If the rock mass is non-uniform and has weak discontinuities in the bolting section, any ground movement along a discontinuity will change the existing tension distribution along the bolt (Zou 2004).

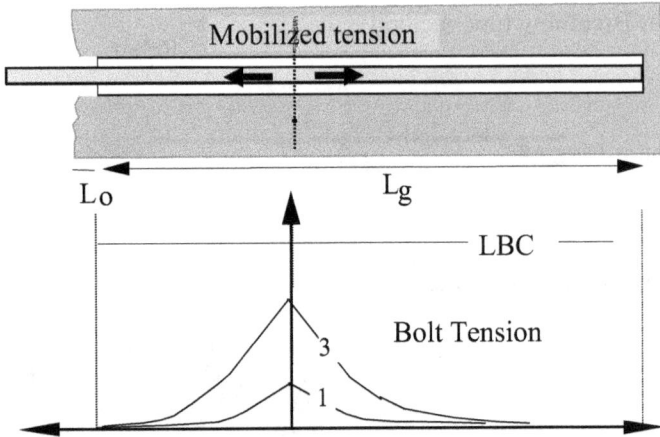

Fig. 9.15 Tension distribution in grouted bolts without bearing plate in a uniform rock mass (After Zou 2004).

d) Rock bolt characteristics and applications

The mechanical characteristics of rock bolts can be determined by laboratory tests on installed specimens. Pull tests indicate that frictional bolts typically demonstrate almost plastic behaviour enduring large deformation, while maintaining original capacity (very similar to that in Fig. 2.3). The capacity of frictional bolts is generally low when compared to others. Fibreglass bolts exhibit linear elastic behaviour, very similar to that in Fig. 2.6 without the OA portion of micro-fracture closure and have higher bearing capacity than steel bolts. Some test results reported by Stillborg (1994) are summarized in Table 9.4.

Rock bolts are widely utilized in the mining industry and have become the primary supporting system. They can be used for a variety of purposes, such as:

- to provide reinforcement by bolting in regular patterns, e.g., 4' x 4' or 6' x 6' spacing, in the roof and/or in the walls,

Table 9.4 Summary of rock bolts test results (After Stillborg 1994).

Bolt type	Bolt size	Typical behaviour	Failure mode	Capacity
Split set	SS39	plastic	ductile	5 tons
Swellex	EXL	plastic	ductile	11 tons
Expansion shell	17.3mm diam.	elasto-plastic	brittle-ductile	9 tons
Grouted steel rebar	22mm diam.	elastic-plastic	ductile	18 tons
Grouted fibreglass	22mm diam.	elastic	brittle	26 tons

- to create a self-supported/reinforced roof beam in stratified rock (Fig. 9.16a),
- to create a self-supported/reinforced arch structure in roof rock around an opening by creating a stress arching effect (Fig. 9.16b),
- to support loose rocks (Fig. 9.16a).

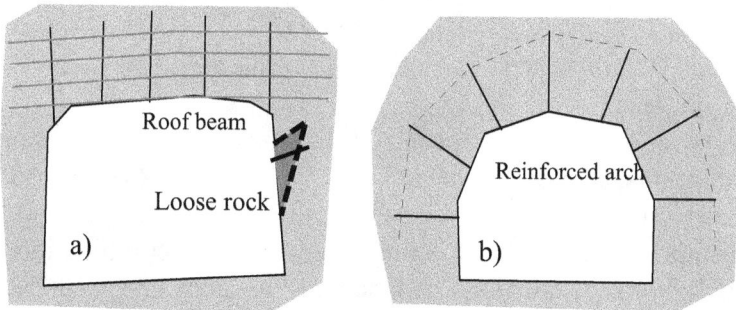

Fig. 9.16 Typical rock bolt applications.

2) Cable bolts

Cable bolts are similar to rock bolts except that cables are longer and somewhat flexible (not as rigid as rock bolts). Cables are made of stranded steel wires and come in a variety of shapes and sizes. Some examples are shown in Fig. 9.17.

There are plenty of publications on cable bolting and its applications. Readers are encouraged to read relevant materials for their application purposes. In general, cable bolts are used for long distance bolting and anchorage, or for permanent support of important underground openings.

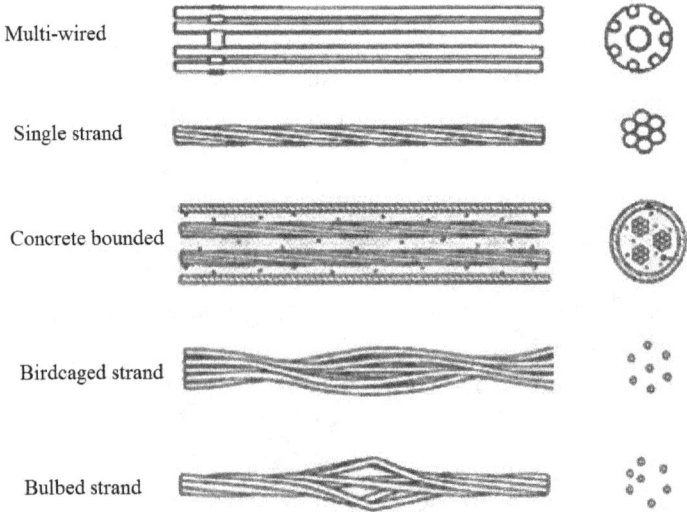

Fig. 9.17 Typical cable bolts configurations.

Cable bolts can be used to make various supporting structures, such as: a) a sling system (Fig. 9.18a) to support loose rock or to reinforce fractured ground over a large span and long distance range, b) cable lacing (Fig. 9.18b), in combination with rock bolts as anchors and screens as secondary support to reinforce fractured ground or ground prone to rock bursts. Cable lacing when combined with screens is good for absorbing shock waves from rockbursts and for permanent openings.

3) Other types of supporting systems

In addition to rock bolts and cable bolts, there are a variety of other types of ground supporting systems, including timber, shotcrete, screens, etc. These will not be discussed in detail in this book. In the following section, only shotcrete and screens will be briefly discussed.

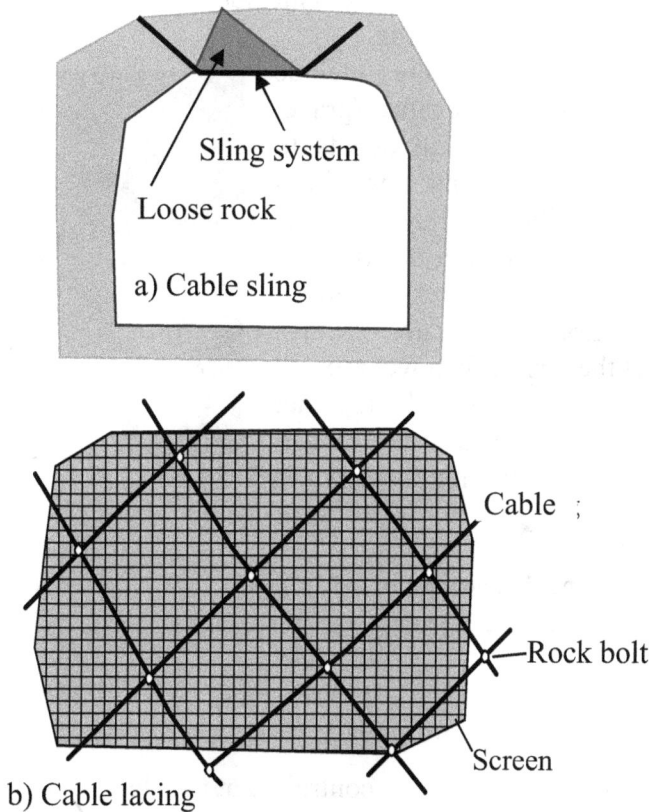

Sling system

Loose rock

a) Cable sling

Cable ;

Rock bolt

Screen

b) Cable lacing

Fig. 9.18 Examples of cable bolt applications. a) vertical section view, b) view looking at opening surface.

a) Shotcrete

Shotcrete has been widely used in geotechnical engineering projects in relatively shallow excavations and surface structures. It is getting more attention in mining operations as well.

Shotcrete is very similar to concrete in its make-up. The primary difference is the proportion of mixing materials, particularly water content. Shotcrete mix is normally loose and able to be delivered with a pumping system via pipes/hoses. In application, shotcrete is sprayed onto the rock surface of underground openings or surface structures through a nozzle under high pressure. The final product looks almost like concrete except the "rough" finishing.

Shotcrete ingredients include cement, water, aggregates (gravel and/or sands), chemical agents to speed up setting, and sometimes reinforcing fibres. They may be mixed dry or wet. In a dry mix, all ingredients except water are mixed together first, and water is then added at the nozzle. In a wet mix, all ingredients are mixed together before being pumped to the nozzle. A typical shotcrete mix is shown in Table 9.5.

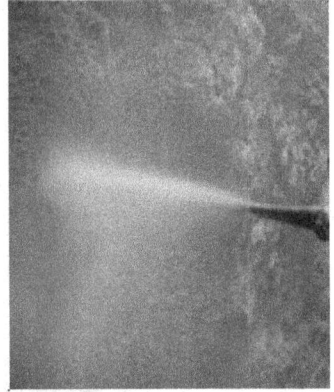

Table 9.5 A typical shotcrete mix.

Ingredients	dry mix	wet mix
Cement	19% by mass	18%
Aggregates	76%	69%
Water	controlled at nozzle	8%
Others (as required)		

Different types of shotcrete machines are available for different applications. In general, a shotcrete machine includes a mixing chamber and a delivery pumping system. In most cases, these are mounted on a mobile vehicle.

Shotcrete often contains reinforcing steel fibres or polymer fibres to increase its flexure ability. The addition of reinforcing fibres has proven to increase both the tensile strength and flexural strength. Kompen (1989) reported that shotcrete reinforced with steel fibre had up to 80% extra strength and the behaviour was more like plastic, but at a reduced capacity as deformation continued.

Shotcrete may be used alone. In mine application, it is usually used in combination with screens, and sometimes with rock bolts or cable bolts as well. Shotcrete is a good alternative supporting system in many applications. However, it is rigid like concrete and not good for burst-prone ground because it lacks the ability to accommodate large deformation and high stress changes.

b) Screens

Screens are often referred to as wire meshes or weld meshes. They are made of single or double steel wires welded together in one inch or larger spacing. Screens are used as secondary supporting systems and are usually applied in combination with rock bolts, cable bolts, and/or shotcrete as illustrated in Figs. 9.18 and 9.19. They are ideal for holding small rock pieces in place, restraining fractured rocks, and also absorbing shock energy released from rockbursts.

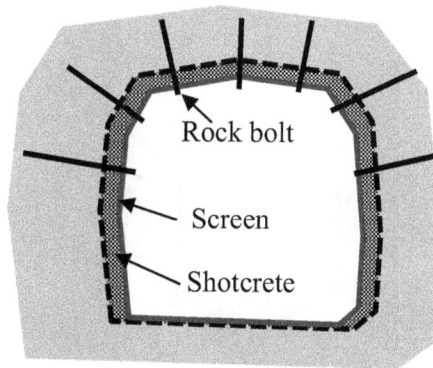

Fig. 9.19 Combined supports.

c) Backfill

Mine backfill, excluding cut and fill mining, here means filling the void (mined stopes) afterwards (Fig. 9.20) and is normally part of mining operations for disposing waste rocks. It is used in open stope mining to prevent the walls from failing by filling up the emptied space with solid material. Common backfill material include waste rocks from underground development, sand excavated on the surface and tailings from processing plants, etc. Backfill is also useful in control of surface subsidence caused by underground excavations.

Since mine backfill is placed after a stope has been completely emptied or mucking completed, at that stage, large ground deformation due to excavation of that stope has already taken place. In comparison to the supporting systems discussed before, mine backfill as ground support simply prevents the rock mass from free movment, rather than reinforcing the rock mass. Such support is referred to as passive support. Passive support will not change what has already happened but will help prevent further deformation due to excavations in adjacent stopes.

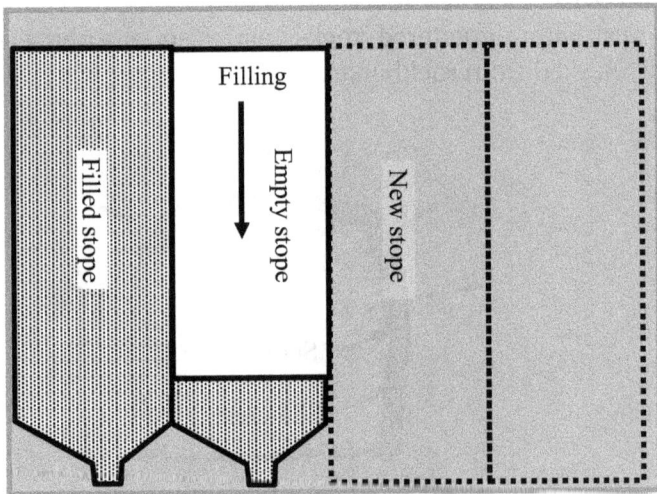

Fig. 9.20 Backfill in open stoping.

9.5 Pre-Reinforcement of Rock Masses

So far, our discussions are focused on the type of supports that are installed after excavation. That means, when the support is installed, fracturing has started and some deformation already exists. In other words, some damage has already occurred to the rock mass around the excavation area. This strategy works fine for most applications.

However, in some cases, it is not desirable to allow much deformation or deformation should be limited to a minimum or as much as possible. Examples include draw points, hanging wall of an inclined stope, stope back of soft orebodies, large major underground structures, etc. In those cases, effort should be made to limit ground deformation by providing support and reinforcement to the ground as early as possible after excavation and sometimes even prior to excavation. This type of pre-support is very effective and useful in mining engineering.

It is realized in many cases that there is no physical access to the place of interest until we make our way by excavation. In some cases, however, pre-reinforcement is possible if careful planning is made in advance. A few examples are discussed in the following.

a) Pre-support of large chambers

If a large chamber is to be excavated in a jointed rock mass, the stability of the chamber can be increased by bolting in advance from neighboring accesses. The example shown in Fig. 9.21 demonstrates bench advance for a large chamber, and cable bolting in advance through an access drift (which in this case is developed specifically for this purpose).

In such a case, if an access drift is not available, the cost of excavating this drift has to be considered.

b) Pre-reinforcement of draw point

The draw point is among the most important areas in a stope. It is the location where the broken ore of the entire stope is to be extracted from. The brow area is under continuous pounding of dynamic forces from the falling ore. Its stability cannot be over emphasized. Patterned and intersecting bolts prior to stope blasting (Fig. 9.22) are very effective in reinforcing the area.

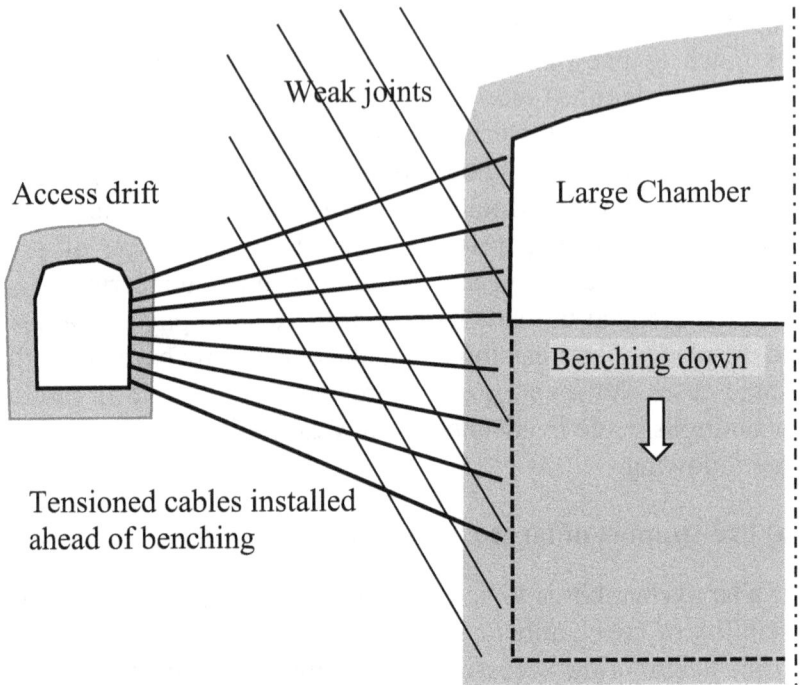

Fig. 9.21 Pre-support of chamber wall.

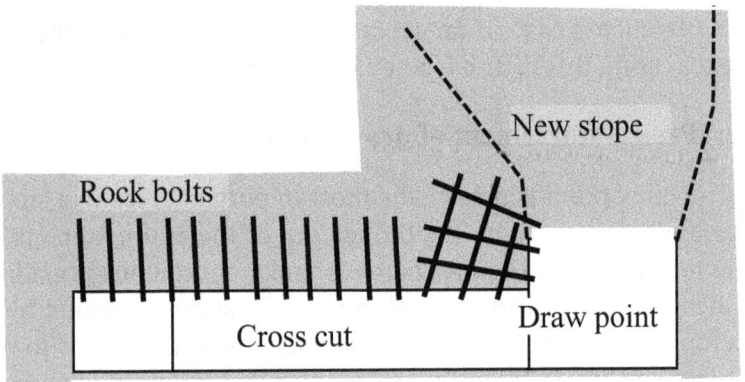

Fig. 9.22 Pre-support of draw point and cross cut.

c) Pre-reinforcement of stope back in cut & fill mining

Cut & fill mining is normally used in weak rocks and orebodies. In this type of mining, the stope is the working place. The stability of the stope back and walls is a major concern for miners' safety. If the back is unable to stand up long enough to allow work to complete in each lift blasting, pre-installed long grouted cable bolts are very effective in supporting the back, as shown in Fig. 9.23.

These bolts are usually installed from a drift above the stope on the top level,

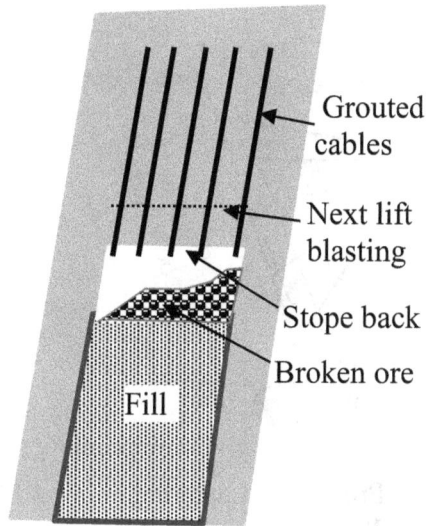

Fig. 9.23 Pre-support to stope back.

and need to be cut off after each round of blasting to clear the space before work can begin for the next lift.

d) Pre-reinforcement of the hanging wall of inclined stopes

Failure of a weak hanging wall or jointed hanging wall is the major source of ore dilution in open stope mining and may also pose a threat to the miner's safety in cut and fill stopes. In open stoping, the inside of the stope is not accessible. If the ground is not strong enough, the hanging wall may fail by spalling, etc., before extraction of ore is completed. It may therefore be necessary to stabilize the hanging wall before mining starts in the stope.

Pre-support to the hanging wall is not an easy task because in most cases there is no access to reach the hanging wall. If an access drift is available, pre-installed cable bolts in the hanging wall will help reinforce the rock and avoid failure. It will depend on the specific condition to provide such support. Figure 9.24 illustrates two cases: an open stope where an access drift is available in the hanging wall, and a cut and fill stope where access is available inside the stope.

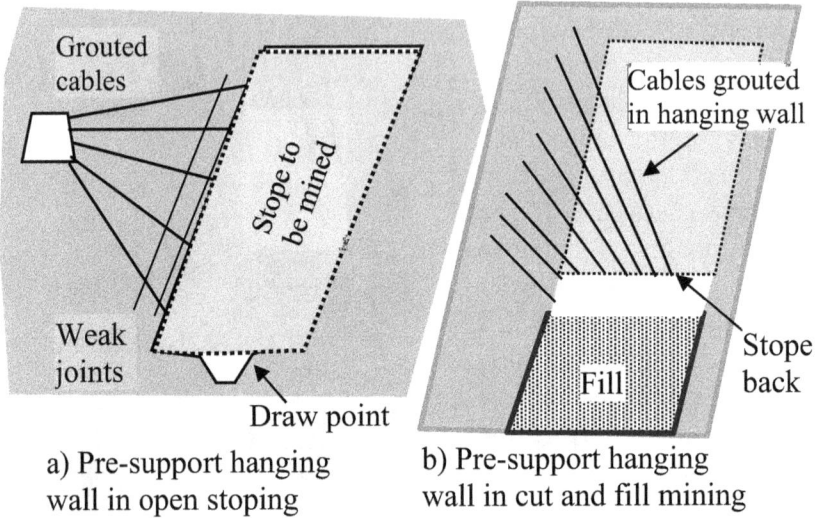

a) Pre-support hanging wall in open stoping

b) Pre-support hanging wall in cut and fill mining

Fig. 9.24 Pre-support of stope hanging walls. a) cables are only grouted in the area near the hanging wall to control the weak joints, b) cables are only grouted in the hanging wall portoin.

Chapter 10

Support of Discontinuity-
Controlled Ground Failure

In this chapter, we will focus on another type of ground failure caused by geological discontinuities, such as faults, joints, bedding planes, etc. These discontinuities cut the rock mass into blocks or wedges, which are held together by the in-situ stresses and the surrounding rock mass. Discontinuities have very low shear strength along their planes and are assumed to have zero tensile strength in the direction perpendicular to their planes. As a result, once an underground opening is created, some of these blocks are exposed and they intend to move to the open space under the stresses in the ground. Sometimes they may fall under the in-situ stress and their own gravity. In the following, we will learn how to identify rock wedges (completely exposed to an opening or hidden) and analyze their stability.

10.1 Graphic Presentation of Discontinuities

1) Definition of geological terms

A geological discontinuity in the field may stretch for a long distance or only exist within a short range. It may have undulation at a large scale. For simplicity, a discontinuity is assumed to be a plane in a three-dimension space. In engineering applications, a discontinuity plane will be defined by its dip direction and dip angle as shown in Fig. 10.1. The dip direction and strike are traced on the horizontal plane. It should be noted that geologists may prefer to use strike and dip to define a discontinuity.

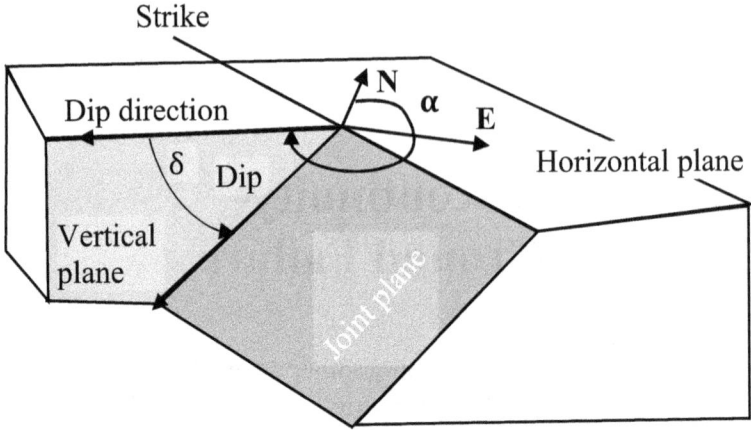

Fig. 10.1 Definition of geological terms.

The strike and the dip direction are perpendicular to each other (Fig. 10.1). The dip direction is defined by the azimuth, α, and is measured in the horizontal plane from the North, clockwise. For example, if a joint dips to the east, its dip direction is 90°. The dip angle, δ, is measured within a vertical plane, which passes through the dip direction and is perpendicular to the strike. δ is measured from the horizontal plane downwards.

In this textbook, a joint plane is expressed as dip direction / dip angle. For example, for a joint dipping 30° to the east, it is expressed as 90°/30°. An equivalent expression is strike N-S and dip 30° E. This joint is represented on a horizontal plane as shown below.

In another example of a joint with a strike 30° measured from the north to the east and a dip 20° to SE, it is expressed as 120°/20° in engineering. An equivalent expression is strike N 30° E and dip 20° SE.

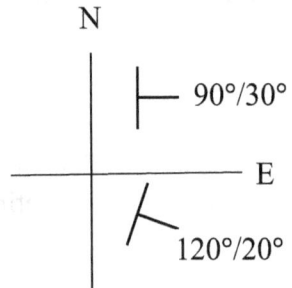

2) Great circles

Graphical presentation of geological data is a very important tool used by geologists and engineers. Use of spherical projections for presentation and analysis of structural geology data has been common practice. In the following, the principles and uses of stereographic projections, or stereo-net, are briefly reviewed.

Imagine a hollow globe being cut in half through the center at mid height, removing the upper half and leaving lower half in place. Then imagine a lamp shining at the position where the top of the original upper half was. The projection of the outer trace of the lower half of the globe on the horizontal plane is a perfect circle. This circle represents a plane with a dip angle of 0°. If the lower half of the globe is virtually cut again through the center but at an inclined angle δ from the horizontal plane (Fig. 10.2a), the projection of the new cutting trace of the globe on the horizontal plane will be an arch, called the "great circle" of the cutting plane (or joint) at dip angle δ (Fig. 10.2b). If the cutting direction is considered, the dip direction of this cutting plane is also defined. On a stereographic net, a joint plane is, therefore represented by a great circle - the trace of the joint plane on the horizontal plane.

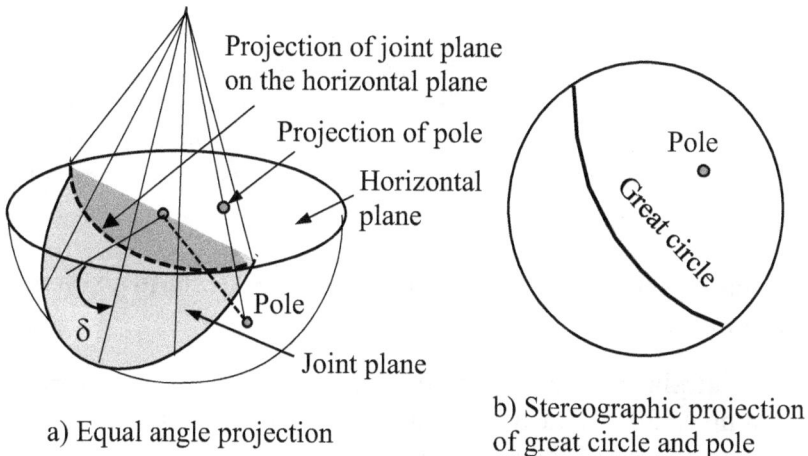

a) Equal angle projection

b) Stereographic projection of great circle and pole

Fig. 10.2 Stereographic projections (After Hoek and Brown 1980a).

In three dimensions, the plane can also be represented by its normal, a vector perpendicular to the plane. As shown in Fig. 10.2a, if the normal of the joint plane is through the center of the globe, it will intersect the other side of the globe's lower half. This intersection point is the "pole" of the plane. On the stereographic net, the projection of the pole is a point, which lies 90° away from the great circle along a diametrical line (Fig. 10.2b).

The dip angle δ of the joint plane in Fig. 10.2a can range from 0° to 90° and the plane can dip to both sides. Therefore, the great circle projection can also be on both sides of the strike line and there will be a 180° difference along the entire diametrical line. When a joint plane is defined by its dip direction and dip angle, there will only be one great circle and one pole.

There are two types of projections: Equal angle projection and Equal area projection. The above illustration is based on the Equal angle projection principle. There is a slight difference in detail in these two types of projections. However, their appearance may not be noticed by general users. In application, if one type of projection is used throughout the process of data presentation and data analysis, there will be no effect on the final results. However, it is not recommended to switch from one type of projection to another in the process. For more detail on the construction of stereographic net, users are suggested to refer to Hoek and Brown (1980a) or other geological books.

Based on the projection principle, stereographic net is generated for processing and presenting geological data. A typical stereo-net is shown in Fig. 10.3. The same stereo-net is used for analysis of underground wedge stability.

Example 10.1 - Construction of a great circle to present a joint plane

A joint plane is defined by its dip direction and dip angle, for example 130°/50°. Construct a great circle and locate its pole on the stereo-net for the joint plane.

Solutions:
(Note in all examples: N = north, E = east, S = south, W = west; other letters are also defined as necessary.)
    Perform the following steps:

- Place a sheet of tracing paper over a stereo-net (use thumbnail to pin them together at the center), trace the whole outer circle and mark N. This will be the reference point to align the tracing paper and the stereo-net in the following steps, as shown in Fig. 10.4.
- Keep N on both papers in alignment and locate 130° on the traced circle as indicated on the stereo-net and mark the location, A.
- Rotate the tracing paper while keeping the stereo-net underneath stable to align the marked point A on the tracing paper with the 90° (or 270°) mark on the stereo-net.
- On the tracing paper, along the E-W (90° - 270°) line count

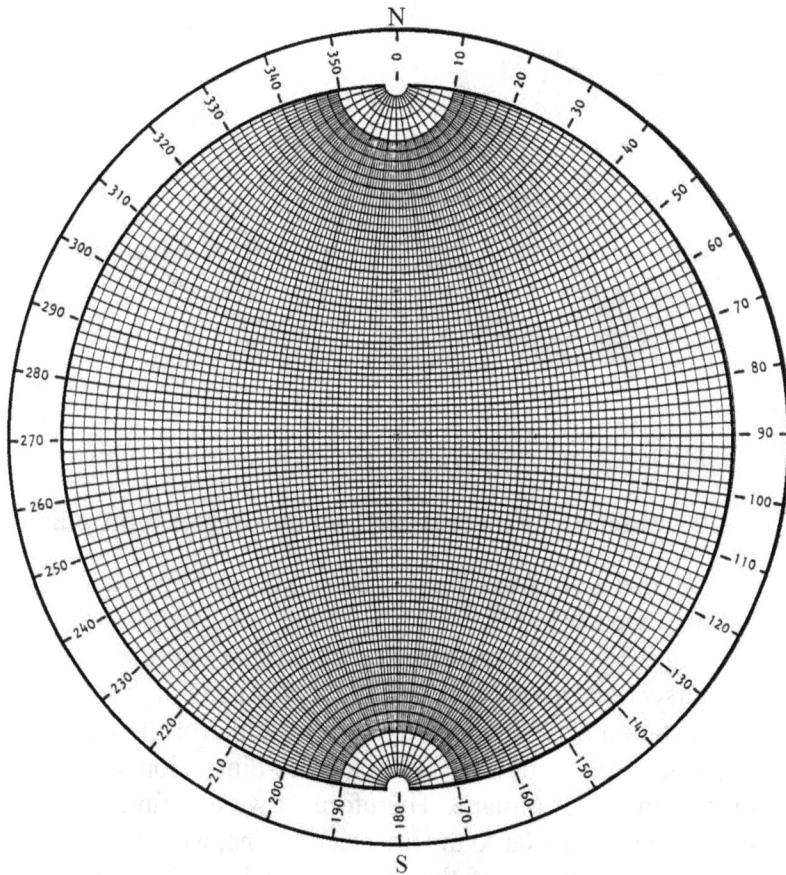

Fig. 10.3 Sample stereographic net.

50° (or five larger divisions) from the edge towards the center. Mark the location, B.

- Keeping the tracing paper in place, trace down the great circle from the stereo-net, which connects the three points: N, B and S (180°). This is the great circle for the joint of 130°/50°.

To locate the pole of the joint plane, keep the tracing paper in place from the last step above. Then along the E-W line, count 90° from B and mark the location C, which will be the pole of the joint plane. If you count beyond the stereo-net edge, continue on the opposite side of the line. The final results are shown on the right in Fig. 10.4.

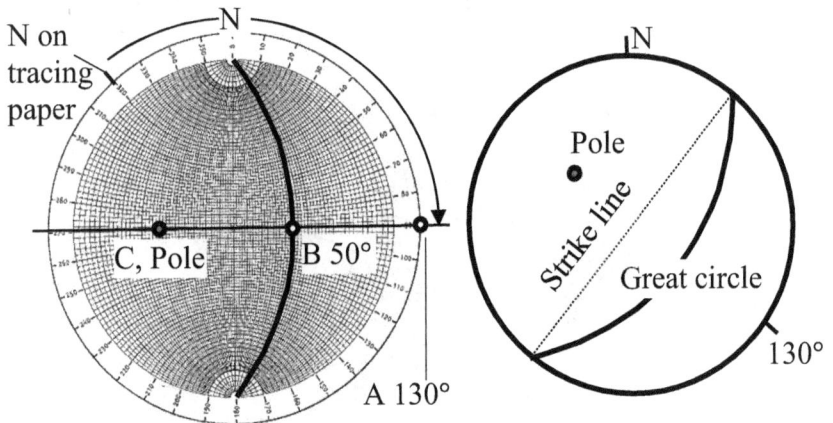

Fig. 10.4 construction of a great circle and location of its pole.

3) Determination of the intersection line of two joint planes

Suppose there are two joint planes intersecting each other, forming a wedge in the rock mass. One of the requirements in assessing the wedge stability is to determine the orientation of the intersection line of the two joints. Similar to a joint plane, the orientation of a line is defined by its trend (or dip direction) and plunge (dip angle). A line in a three-dimension space is in fact the normal of a plane. Therefore, if we can find the plane which is perpendicular to the intersection line, we can determine the trend and plunge of the intersection line. The process is demonstrated in the following example.

Example 10.2 – Determine intersection line orientation

Consider two joint planes, defined as 130°/50° and 250°/30°, respectively. Determine the trend and plunge of the intersection line.

Solutions:

Perform the following steps (Refer to Fig. 10.5):

- Place tracing paper over a stereo-net, trace the outer circle and mark N. Construct two great circles for the two joint planes and locate their poles as discussed before. The two great circles will intersect each other at a point, which represents the intersection line in space.
- Draw a line through the center and the intersection point on the tracing paper. Count from North clockwise to the line to determine the trend of the intersection line. It will be approximately 200° in this example.

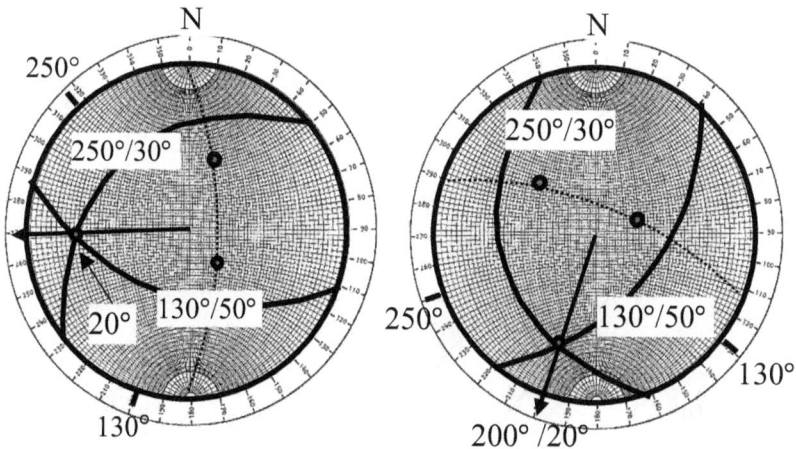

Fig. 10.5 Determination of the intersection line of two planes.

- Rotate the tracing paper over the stereo-net to align the intersection line you just drew with the E-W (Fig. 10.5, left). The two poles of the two joint planes will lie in the same plane. Trace a great circle connecting the two poles from N to S.

- Count the degree on the E-W line from the edge to the intersection point to determine the plunge. It will be about 20° in this case.

The results are shown on the right in Fig. 10.5 and the orientation of the intersection line is 200°/20°.

4) Relationship between true dip and apparent dip

In an underground drift, it is often possible to trace a joint plane on the roof or the sidewalls (Fig. 10.6). If the drift is perpendicular to the joint plane, the traced dip angle on a vertical sidewall, measured from the horizontal line, will be the "true dip" δ of the joint. Otherwise, it will be the "apparent dip", δ', which is smaller than or equal to the true dip, i.e., $\delta' \leq \delta$.

To determine the true dip, we need to know the orientation of the drift (trend and plunge) and the strike or dip direction of the joint plane.

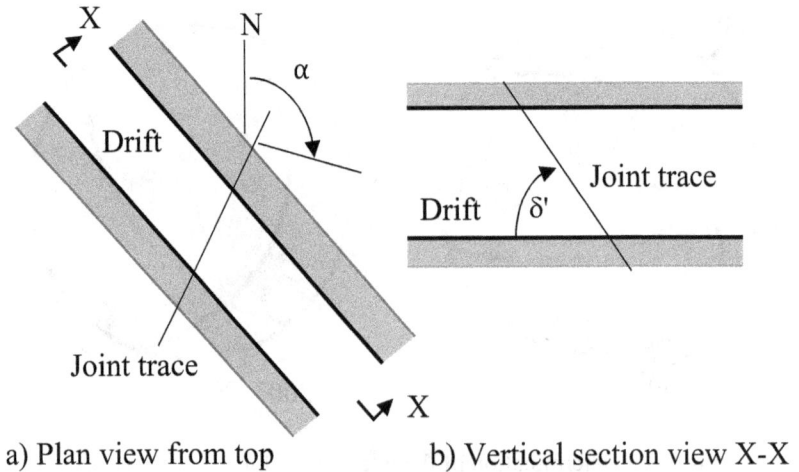

a) Plan view from top   b) Vertical section view X-X

Fig. 10.6 Traced dip angle of a joint on drift wall.

Example 10.3 – Determine true dip from apparent dip

Suppose that a horizontal drift as shown in Fig. 10.6 has a trend 320° – 140°, and a joint plane with dip direction of 110° intersects the drift. If the trace of the joint on the vertical wall has a dip angle of $\delta' = 40°$, what is the true dip of the joint?

Solutions:

From the given condition, it is clear that the joint is not perpendicular to the drift (as indicated in Fig. 10.6a) and the measured dip angle is an apparent dip angle. The vertical wall of the horizontal drift has a dip angle of 90°.

To determine the true dip, follow these steps:

- Place tracing paper over a stereo-net, trace the outer circle and mark N.
- Draw a line for the drift wall. The drift wall can be considered as a plane which has a strike in the same direction as the trend of the drift. The dip direction of this plane can be determined if the dip of the wall is known. In this example, the wall is vertical and the dip direction becomes irrelevant. Draw a straight line through the center from 320° to 140° to represent the drift wall.
  Note: If the wall is not vertical, we need to find the dip direction and dip angle of the wall and then draw a great circle by following the procedure described previously.
- Find the strike of the joint. Since the dip direction of the joint plane is 110°, add and subtract 90° from 110° to get 200° and 20°. Mark these two points on the boundary circle of the tracing paper, which will be where the great circle of the joint intersects the boundary circle.
- Rotate the tracing paper to align the drift with E-W, count 40° from the East edge (because the joint dips to SE) along the drift line, and mark that point.
- Rotate the tracing paper to align the marked two points of 20° and 200° with N and S respectively (Fig. 10.7a).
- Trace a great circle to connect the N, S and the point marked 40° on the drift line.

- Keep the tracing paper in place and count along the E-W line from the edge to the great circle to determine the true dip angle of the joint plane, which is 44° in this example. The result is shown in Fig. 10.7b.

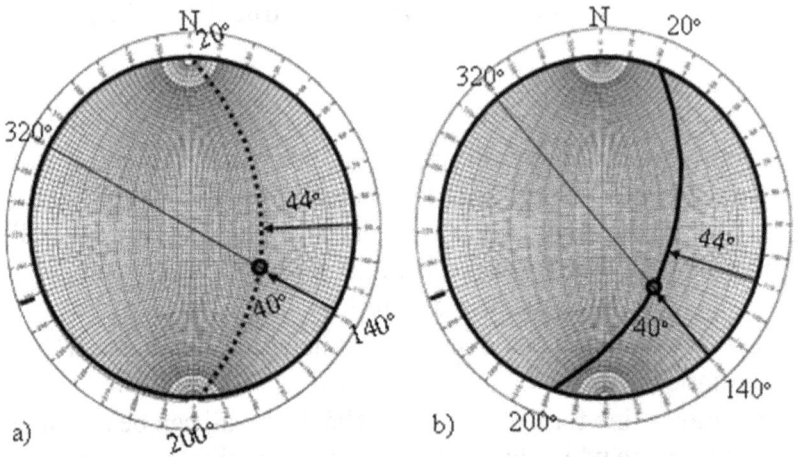

Fig. 10.7 Determination of true dip.

5) Determination of contour of poles for joints

When a large quantity of data of geological structures are collected in the field, it is necessary to analyze the data to identify any trend or clusters. Stereographic net is a useful tool in this task. A pole stereographic projection, similar to the one shown in Fig. 10.3, is very useful for counting and grouping poles. A typical stereo-net is shown in Fig. 10.8.

On the pole stereo-net, instead of plotting a great circle, a pole is plotted. When a large number of joints are mapped, they can be presented by the same number of poles on the pole stereo-net. Use of the net is straight forward and is very similar to the one in Fig. 10.3. A joint plane is identified by its dip direction and dip angle. One needs to keep in mind however, that there are two numbers on the outer boundary in Fig. 10.8: one is for the joint plane and another for its pole. The dip direction of the pole is obtained by adding or subtracting 180° from the dip direction of the joint plane. Therefore, for poles, 0° is at the center and 90° is at the edge of the net.

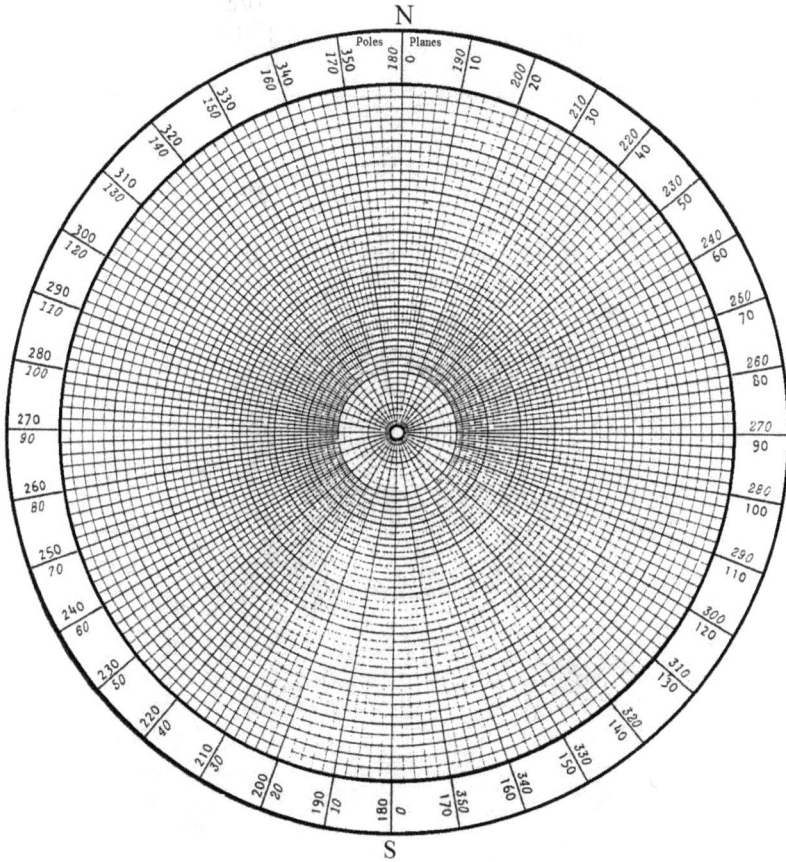

Fig. 10.8 Sample polar stereographic net.

An example of pole plots is shown in Fig. 10.9. A total of 95 poles are plotted. Based on the clustering, these poles belong to four separate sets plus one "outlier" – a fault, which has to be treated separately. For each group, mean values of dip direction and dip angle can be determined from the center of the cluster and the results are shown below.

Set #	Joint plane (°) (dip direction/dip)	Pole (°)
1	350/22	170/22
2	230/80	50/80
3	321/85	141/85
4	80/63	260/63

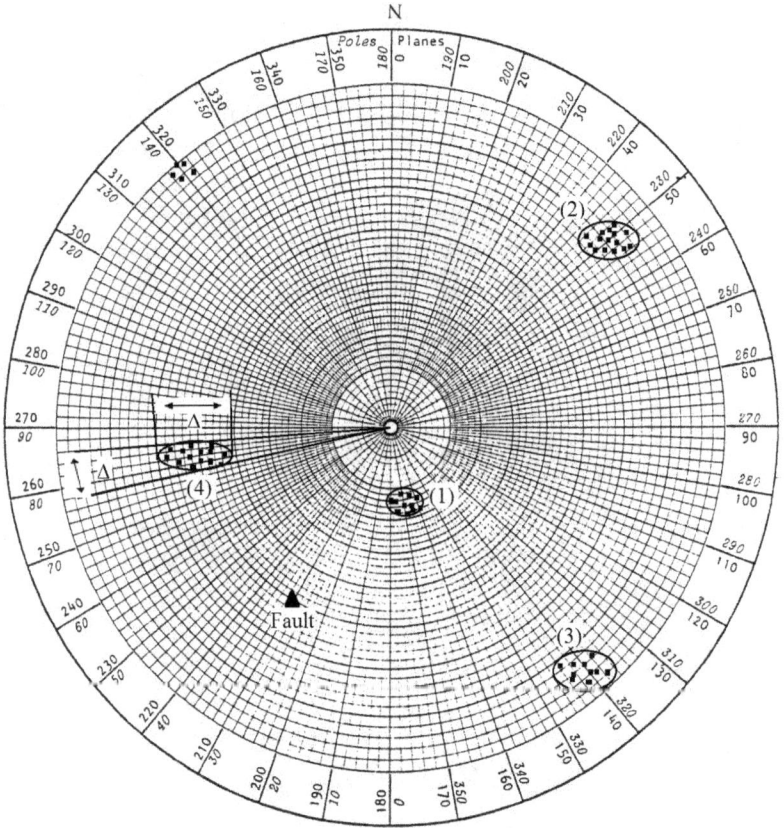

Fig. 10.9 Example of pole plots.

Variation, Δ, of the dip direction and dip can also be estimated as illustrated in Set (4) in Fig. 10.9. It is clear here that this geological data cannot be simply mixed together to calculate an average. They must first be separated into groups before an average can be determined.

It is also important to note that a fault usually has very different properties from joints and will play a significant role by itself in ground instability. It should therefore be separated from other joints and be treated separately; even if it falls within a set of joints. The same principle applies to other types of major geological structures, such as a weak shear zone.

10.2 Underground Wedge Failure Analysis

Discontinuities in the ground separate the rock mass into blocks in various geometry and sizes. The simplest geometry is a wedge with four faces, as shown in Fig. 10.10.

In the underground, not all wedges are a problem. A wedge becomes a concern for instability only when it is exposed to an opening and is likely to move, either by falling or sliding. If a face of a wedge is completely exposed, there is a good chance of failure (dislocating from where it should be). If a wedge is only partially exposed, it may have potential for failure. The key is how to identify an unstable wedge and to estimate its potential for failure.

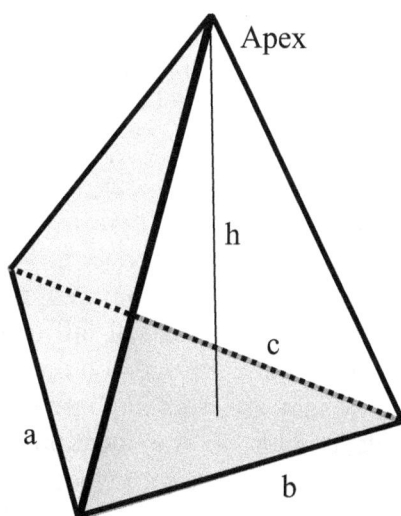

Fig. 10.10 Geometry of a potentially unstable wedge in the roof.

1) Wedge failure analysis with stereo-net

Depending on where the wedge is located, it may or may not pose a problem. Often the weight of a wedge plays a significant role. The most vulnerable location is therefore in the roof, followed by the sidewalls.

a) Roof wedge analysis

In the underground, the opening face (roof or sidewall) provides one free surface. Therefore, to form a wedge underground, only three more non-parallel joint planes are needed. If there are three or more joint planes, any three of them may form a wedge with the roof of an opening (an exposed surface). All combinations need to be considered. Furthermore, a wedge will only be a concern if it sits in a vulnerable position and if one side is completely exposed to the opening. Otherwise, the wedge may still have support from the rock mass in the unexposed portion.

The necessary conditions for roof wedge failure:

Criterion 1: A minimum of three joints must exist to form a wedge underground.

Criterion 2: The whole base of the wedge must be exposed to an opening.

Depending on the shape and position of a wedge, when Criterion 2 is satisfied, a wedge may be free to fall or free to slide. For a roof wedge, this will depend on the location of the gravity center of the wedge. A simple method to determine whether a wedge will fall or slide is to draw a virtual vertical line through the apex on the top of the wedge (Fig. 10.10). This line may intersect the base of the wedge or fall out of the base, depending on the location of the apex. Further criteria for assessing how a roof wedge is going to fail therefore are:

Criterion 3: If the vertical line through the apex falls within the base line in the vertical cross section (Fig. 10.11a), the wedge is free to fall.

Criterion 4: If the vertical line through the apex falls out of the base line in the vertical cross section (Fig. 10.12a), the wedge may be free to slide, either along one of the joint planes or one of the intersection lines of the wedge, provided that the shear force exceeds the shear resistance.

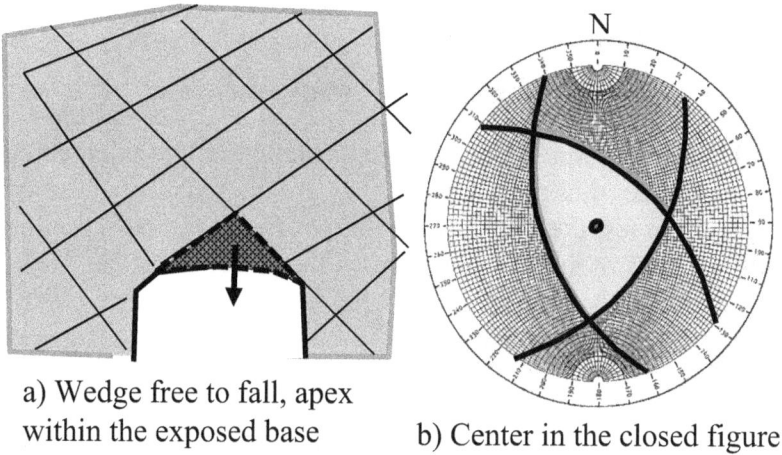

a) Wedge free to fall, apex within the exposed base

b) Center in the closed figure

Fig. 10.11 Conditions for a wedge free to fall.

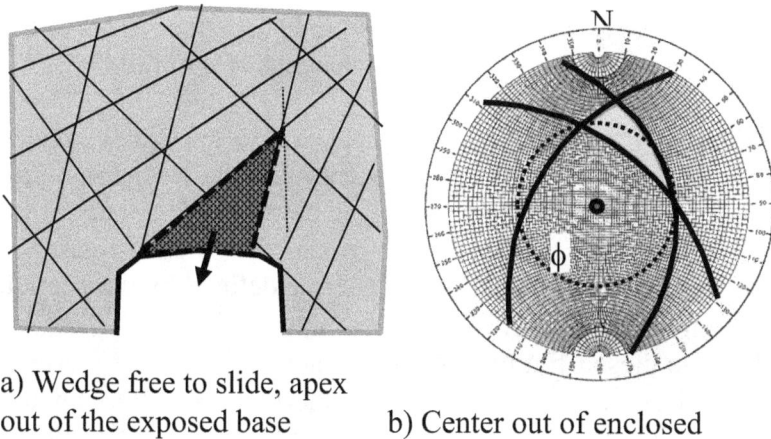

a) Wedge free to slide, apex out of the exposed base

b) Center out of enclosed

Fig. 10.12 Conditions for a wedge free to slide. The dotted circle in b) represents the frictional angle ϕ of the sliding plane or intersection line.

It is not possible to draw such a vertical line in the field since a wedge is hidden in the rock mass above the roof. The only thing visible from inside an underground opening is the exposed triangular base of the wedge. Another method is required to accomplish this task. In addition, methods are needed to evaluate the conditions specified in Criterion 4.

b) Use of a stereographic net for wedge analysis

With a stereo-net, the Criteria 2 to 4 can be easily implemented. For any three joint planes, when they are plotted on a stereo-net as great circles, they form some kind of figure on the stereo-net with the apex of the wedge located at the center of the net. Based on the formed figure, assessment can be made on wedge stability. The conditions for assessing unstable roof wedges on a stereo-net are:

Criterion 5: If the three great circles of three joints form a "closed figure" on the stereo-net (i.e., they all intersect each other), the three joints form a wedge and the base of the wedge is exposed.

Criterion 6: If the "closed figure" surrounds the center of the stereo-net (Fig. 10.11b), the wedge is free to fall.

Criterion 7*: If the "closed figure" falls to one side and does not cover the center of the stereo-net, the wedge may fail by sliding. Sliding takes place along a joint plane or an intersection line if at least part of the "closed figure" is within the frictional (dotted) circle defined by the friction angle ϕ (Fig. 10.12b).

When the closed figure falls inside the frictional circle, the dip angle θ of the plane or the intersection line is steeper than its frictional angle, as illustrated in the margin diagram.

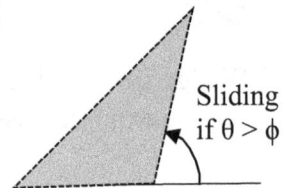

Sliding if $\theta > \phi$

* For Criterion 7: Based on the gravity of the wedge only, if the "closed figure" falls completely out of the frictional circle, the wedge is stable because the dip angle of the sliding plane or intersection line is smaller than the frictional angle ϕ. However, it must be kept in mind that if there are high stresses in the ground, even the "closed figure" falls out of the frictional circle, sliding failure may still be possible due to in-situ stresses.

c) Determination of wedge volume

Volume of a wedge is calculated by

$$V = h \, A_b/3 \qquad (10.1)$$

where h is the height from the apex to the base (Fig. 10.10) and A_b is the area of the base of the wedge, defined as

$$A_b = \sqrt{s(s-a)(s-b)(s-c)} \qquad (10.2)$$

where $s = (a+b+c)/2$ $\qquad (10.3)$

The base dimension of a wedge can be measured underground by tracing the exposed area. However, the height h has to be determined by other means based on the joint planes and the base of the wedge.

Assume that three joints, A, B and C, identified by their dip directions and dip angles form a wedge. The corresponding lengths, a, b and c, on the base of the wedge are measured from the exposed traces on the roof of an underground drift. To determine the height of the wedge, follow these steps:

- Place tracing paper over a stereo-net, trace the outer circle and mark N.
- Construct three great circles on the tracing paper for the three joints (A, B, C) and mark their strike lines (a, b, c) by connecting the two ends of each great circle (Fig. 10.13a).
- Draw lines of intersection (ab, bc, ac) by connecting the center to the intersection points between each pair of great circles as indicated (Fig. 10.13a).
- Draw a true plan view for the roof of the drift in a rectangular shape, with width d proportional to the width of the drift by scale and length along the trend of the drift (Fig. 10.13b).
- On the plan view in Fig. 10.13b, draw three lines (a, b, c) to be parallel to the a, b, c lines in Fig. 10.13a, respectively, forming the base of the wedge. Pay attention to the dip directions of A, B and C to decide proper positions for a, b and c.
- On the plan view in Fig. 10.13b, draw three intersection lines ab, bc and ac to be parallel to the ab, bc and ac lines in Fig. 10.13a, respectively. They should intersect at the apex.

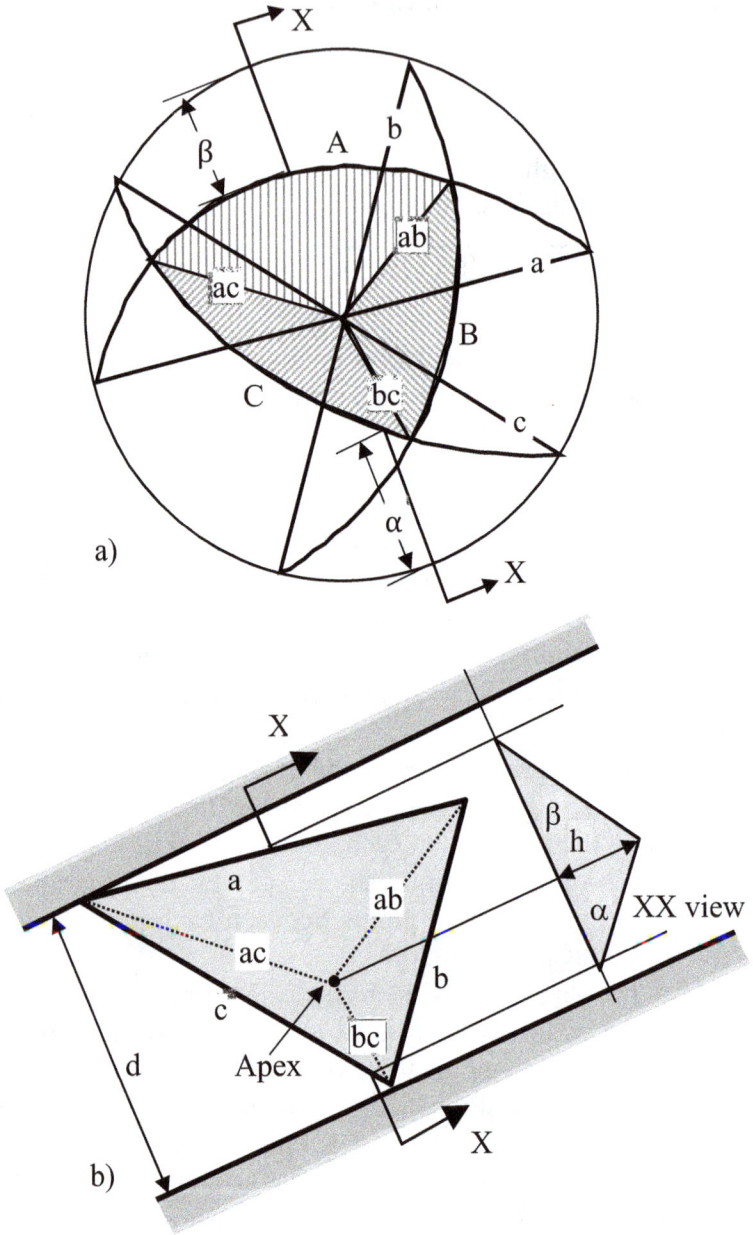

Fig. 10.13 Use of stereographic net for determination of the shape and
height of an underground wedge formed by joints.

- On the plan view in Fig. 10.13b, draw a line XX through the apex, perpendicular to the drift axis. The XX section will be the vertical section through the apex.

- On the stereo-net trace in Fig. 10.13a, draw a line XX through the center and parallel to the XX in Fig. 13b. Mark the intersection points between XX and great circles A and C.

- Rotate the tracing paper to align the XX line with E-W and count to determine the apparent dip angles β and α for A and C, respectively.

- Go back to the plan view in Fig. 10.13b, draw two lines parallel to the drift axis through the intersection points between XX and the base lines a and c. On the right side, draw a line perpendicular to those two lines you just drew as the base line of the roof. From the base line, starting from its intersection points draw two lines with angles β and α to complete the vertical section view of the wedge. The intersection of the last two lines is the actual apex. Measure the distance h as shown in Fig. 10.13b and convert h to wedge height according to the scale you used.

d) Sidewall wedge analysis

Since the sidewalls are vertical or nearly vertical and the stereographic projection is on a horizontal plane, the stereo-net cannot be used directly for sidewall wedges. To resolve this issue, a wedge in the sidewalls needs to be transformed. In other words, the projection of joint planes on a horizontal plane needs to be transferred to the projection on a vertical plane. To demonstrate the process, let us look at the following example.

Consider a drift with a trend direction from 70° to 250° in a rock mass where three joint sets are present. These joints are represented by the great circles marked A, B and C in Fig. 10.14a. These great circles are the projection of these joints onto a horizontal plane and the closed figure represents what a wedge in the roof would look like. In order to find the shape of the wedge in the sidewall, it is necessary to determine the shape of the closed figure projected onto a vertical plane. This is done by transformation using stereo-net as follows:

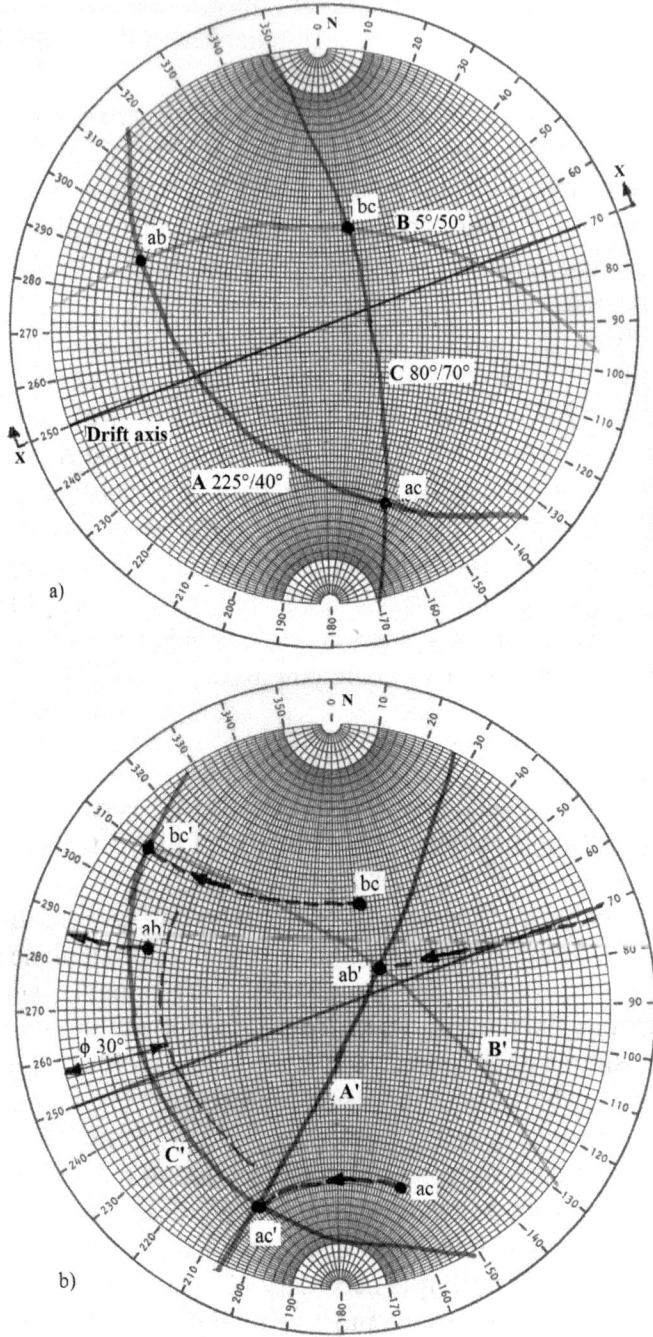

Fig. 10.14 Transformation of wedge projection from a) horizontal plane to b) vertical plane.

To determine the closed figure of the wedge in the sidewall, the great circle intersections ab, bc and ac are rotated by 90° about the drift axis by following these steps:

- Trace the points, ab, bc and ac, onto a clean piece of tracing paper. Trace the outer most circle and mark the centre, N and also the drift axis on this tracing.
- Locate the tracing paper on the stereo-net, using a centre pin so that the outer most circle and the North point coincides with those on the net (Fig. 10.14b).
- Rotate point ab by counting off 90° to the left along the *small circle* passing through the point ab (the small circles are perpendicular to the great circles on the stereo-net). It should be noted that the small circle through ab passes out of the net circumference and re-enters the net at a diametrically opposite point. Continue counting and mark the end point as ab'.
- Do the same for points bc and ac to locate points bc' and ac'. It should be noted that the rotation of all the points must be in the same direction, left or right. This procedure will ensure that all intersection points lie within the same hemisphere and that the projection onto the vertical plane is correct.
- Rotate the tracing paper left or right around the center, to align a great circle with each pair of the new intersection points, ab', ac' and bc'. Trace down the three great circles A', B' and C'. The projections of the joint planes on the vertical sidewall and the closed figure are shown in Fig. 10.14b.

The strike lines of these great circles (i.e., the straight lines connecting the two ends of the great circles, a', b', and c', not shown in Fig. 10.14b) represent the traces of the joint planes on the vertical sidewalls of the drift. Construction of the true view of the wedge in the sidewall follows the same procedure as that for the roof.

2) Use of computer software for wedge failure analysis

As demonstrated above, it is a tedious process to conduct wedge stability analysis using the stereo-net approach by hands. The size and shape of potential wedges in the rock mass surrounding an underground opening depends on the size, shape and orientation of the opening and also on the orientation of the discontinuity sets. The three-dimensional geometry of the problem necessitates relatively tedious calculations. It would be much more efficient to use specialized computer software for the task.

There are commercially available computer software packages for underground wedge analysis. Some may be free. The author is not intending to endorse any particular brand here. Individual users may find one of personal preference. Use of computer software is straightforward. With a short period of training, one should be able to master the how-to skills. What is important is that a user must know what the computer software is doing and be able to make judgement on whether the results generated by the software are meaningful. All of the above discussions in this chapter are intended to prepare readers for this task.

In general, the required input data would include orientations of the joint planes and their shear strength property parameters, size, shape and orientation of underground openings. The output results would include plots of great circles, 2D and/or 3D wedge diagrams and their locations in the rock mass around the opening. The stability of a wedge may be presented and sometimes options to provide support may also be included.

In a rock mechanics course, software should be acquired or provided by the instructor and detail of using the software should be discussed accordingly.

10.3 Underground Support of Loosened Rocks

Depending on the shape and position of a wedge, the support strategy will be different. The idea is to provide sufficient force to keep the wedge in place, either by preventing the wedge from sliding or by anchoring it to the rock mass deep inside to increase the overall integrity.

Rock bolts and cable bolts are the most common primary supporting systems used to support wedges in underground openings. Cable bolts provide a longer reach and are intended for larger wedges.

a) Support of free fall wedges

If a roof wedge is free to fall as shown Fig. 10.15, it moves independently from the rest of the rock mass and applies concentrated or eccentric loading to the support system.

A support needs to provide a supporting force sufficiently to carry the whole weight of the wedge. Rock bolts and cables work best in this case to resist this type of load, while steel sets and concrete lining should be avoided. The minimum number of rock bolts needed is calculated by the following equation:

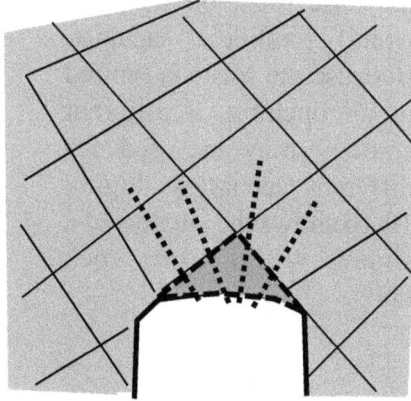

Fig 10.15 Support of a free-fall wedge in the roof.

$$N_{min} = W / T \qquad (10.4)$$

where W is the weight of the wedge calculated based on its volume and gravity, and T is the capacity of tension in a bolt.

The actual number of bolts will depend on the safety factor and the effectiveness of bolting. The effectiveness of bolting will depend on the location where a bolt is installed relative to the wedge and the depth of anchoring in the rock mass. It can be imagined that if a short bolt is installed inside the wedge to be supported or installed near the tip of the wedge, the bolt may carry little or no load at all and becomes ineffective.

An effective bolt should have proper length to pass well beyond the wedge boundaries and reach the stable rock mass behind, as illustrated in Fig. 10.15.

b) Support of roof wedges free to slide

If a roof wedge is free to slide by its own weight, the support system does not need to carry the entire weight of the wedge, but only needs to provide a force sufficiently to resist it from sliding.

This may be achieved by anchoring the wedge to the surrounding rock mass or to the base rock, on which it would slide against. Because the space is limited in an opening, a different strategy may be used for different scenarios. Figure 10.16 shows the support of a sliding wedge using rock bolts.

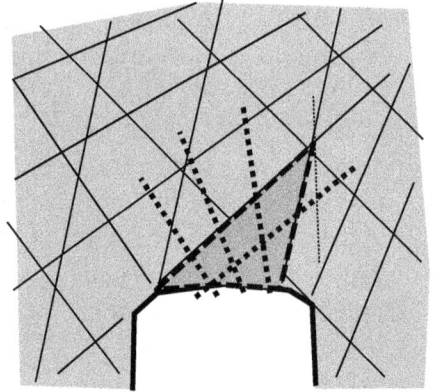

Fig 10.16 Support of a sliding wedge in the roof.

c) Support of wedges in sidewalls free to slide

If a wedge is identified in sidewalls, its only possibility of failure is to slide on its base or an intersection line formed by two joints. In this case, the effective supporting strategy is to pin the wedge to the base rock using rock bolts or cables, as shown in Fig. 10.17. An effective rock bolt should be installed at a proper angle towards the sliding base. The supporting force should satisfy the following condition:

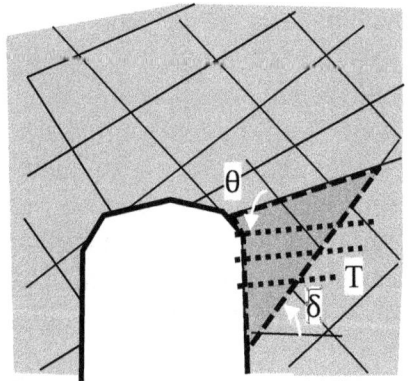

Fig 10.17 Support of a sliding wedge in the sidewall.

$$R \geq SF \times F \tag{10.5}$$

where R is the total resistance force generated by all rock bolts and F the sliding force.

Consider a case where the installation angle between the rock bolts, which are installed in parallel, and the normal of the sliding base (a plane) is θ and the dip angle of the sliding base is δ (Fig. 10.17). Based on the equilibrium condition, a safety factor is given by

$$SF = \frac{R}{F} = \frac{cA + (W \cos\delta + T_t \cos\theta) \tan\varphi}{(W \sin\delta - T_t \sin\theta)} \tag{10.6}$$

where A is the sliding base area of the wedge and T_t is the total of the minimum tension force from all rock bolts. By rearranging Eqn. 10.6, the minimum tension force is determined by

$$T_t = \frac{W (SF \sin\delta - \cos\delta \tan\varphi) - cA}{\cos\theta \tan\varphi + SF \sin\theta} \tag{10.7}$$

and the minimum number of bolts required is

$$N = T_t / T \tag{10.8}$$

where N is to be rounded up to the nearest higher whole number.

A safety factor of 1.5 to 2.0 has been recommended by Hoek and Brown (1980a), depending on the damage which would result from sliding of the wedge and whether or not the bolts are grouted.

When the geometry of a wedge is in such a position that it would cause sliding along a line of intersection of two joints, the above analysis can be used to give a first approximation of the required support load and number of bolts. This is expected to give a conservative answer and the estimated tension load in bolts would be higher than that actually required. In many practical applications, this would be acceptable since the extra work involved and the cost savings in rock bolts would not justify a more detailed analysis. However, in the case of very large underground caverns, the sizes of wedges or blocks can be considerable and a more precise analysis may be required.

d) Support of wedges with shotcrete

Shotcrete may be used as additional support for wedges in blocky ground and can be very effective if applied correctly, particularly for smaller wedges and blocks. The base of a typical wedge has a large perimeter and hence, even for a relatively thin layer of shotcrete, a large cross-sectional area of the shotcrete has to be punched through before the wedge can fail.

Consider an example as shown in Fig. 10.15. The base of a roof wedge across the whole span of a 5 m drift could have a perimeter of 15 m or more. A layer of shotcrete 50 mm thick will have a total cross-sectional area of 0.75 m^2 available to provide support for the wedge. Assuming a shear strength of 0.34 MPa (34 tons/m^2, see Table 9.1) for the shotcrete layer, it would be able to support a wedge weighing 25.5 tons.

It is important to ensure the quality of the shotcrete which should be well bonded to the rock surface in order to prevent a reduction in support capacity due to peeling-off of the shotcrete layer. If the shotcrete is not bonded to the rock surface well, when a wedge moves, it will exert a "point" load or "line" load on the shotcrete. In this situation, the shotcrete will then be in a bending/flexure condition and the supporting capacity will be lower.

Another problem of using shotcrete to support wedges is that it has very little strength at the time of application and it takes several days to reach its full strength. When wedges require immediate support, the use of shotcrete for stabilisation is clearly inappropriate. However, if rock bolts are used to ensure the short term stability of the rock mass, a layer of shotcrete will provide additional long term stability.

It is believed wasteful to use shotcrete in strong rocks with large wedges as only the shotcrete covering the exposed perimeter of the wedge is effective in providing any resistance. The ideal application for shotcrete is in highly jointed rock masses, where wedge failure would occur as a progressive process, starting with smaller wedges exposed at the opening surface and gradually working its way into the rock mass. In this situation, shotcrete provides very effective support and deserves special attention.

Chapter 11

Numerical Modelling Application in Mining and Geomechanics

11. 1 Concept of Numerical Modelling

In underground engineering projects, such as mine excavations, the stress-displacement relationships are very complicated due to many factors: the geological structure, variation of rock formation, opening shape, interaction and mutual impacts of multiple openings, etc. Only in a very simple case of a circular opening, can we calculate the changes of stress and deformation. In other cases, it is beyond our imagination. Numerical modelling in these situations becomes a very useful and versatile tool. It has been commonly used in engineering practice to simulate excavations in rock masses as well as determine the new stress distribution and displacement in the vicinity of excavations.

In this chapter, we will learn the fundamentals of a few commonly used numerical modelling methods and their applications in mining and geomechanics. In particular, we will answer questions like: what is numerical modelling? what are the modelling principles? how does numerical modelling work? how to use it in practice? However, this chapter is only intended as an introduction. Detailed description of each method is beyond the scope of this book. Users are suggested to read relevant books and publications for more technical detail.

Meaning of modelling in different applications

Numerical modelling has different meanings in varying applications. Although the same principle is used, it may mean a completely different thing in another field. Modelling generally means to simulate or to reproduce certain processes under specific conditions. For example, in groundwater movement analysis, the process involves water movement from a high-pressure point to a low-pressure point in the ground and the movement is governed by the relationship between pore pressure and hydraulic conductivity of the ground. Oil and gas movement in a reservoir formation follows the same principle as groundwater movement. In these situations, actual physical relocation of a substance from point to point occurs. A numerical model will simulate how water or oil moves in the ground.

In heat transfer analysis, the process involves heat transfer from a high temperature point to a low temperature point in a medium. The transfer is governed by the relationship between the heat source and thermo conductivity of the medium. In this situation, no physical relocation of a substance occurs but the effect of heat is "spread" to the surrounding locations. A model will simulate how heat is transferred in the medium.

In geomechanics applications, what we are interested in is the effect of excavation in the ground in terms of stress and displacement. Excavation does not cause physical relocation of rock like in groundwater movement but causes changes of stresses in the surrounding rock and in turn causes deformation in various magnitudes. The process to be analysed is the stress change and displacement in a rock mass and the process is governed by the stress-strain/deformation relationships in the rock and the rock properties. A numerical model will simulate how much stress is changed and how much deformation occurs after excavation in the surrounding rock.

There are many more such examples. Each is different.

Physical simulation versus numerical simulation

Numerical modelling is often referred to as numerical simulation. Let's first start with an example of physical simulation in a laboratory. In general, a physical model is built, often on a smaller scale, based on certain criteria to simulate a real process in nature. We have all ridden or seen a bus in our life and some may also have ridden a high-speed train. The former looks like a rectangular box and the latter like a bullet head. You may have wondered why these two types of vehicles are so different in shape. You may say because the bullet head shape has less resistance to air and is more efficient at higher speed. That is true, but how was that conclusion made? We can build a wind tunnel with the capability of changing air flow as well as small model vehicles in various shapes, put the model vehicles in the wind tunnel and test them at various wind speeds. By monitoring air velocity and pressure at various locations of the model vehicle under different wind speeds, we can collect and analyse relevant data and eventually determine the optimal shape for the vehicle. What we see today is the outcome of many experiments (or physical simulations).

However, in many cases, conducting a physical simulation is either too complicated, impossible due to various reasons, or simply too expensive. If the involved factors and parameters of a process are known to have certain correlations, or an explicit mathematical formula, the process may be simulated numerically without physical simulation.

To better explain the concept, let us look at a simple example – a spring model as shown in Fig. 11.1. At a given magnitude of the applied force F, the displacement δ can be measured in a laboratory test, a physical simulation of the spring behaviour under pulling force.

As discussed in Chapter 2, the reaction of this model to a pull force can be described by the following relationship

$$F = k\delta \qquad\qquad (11.1)$$

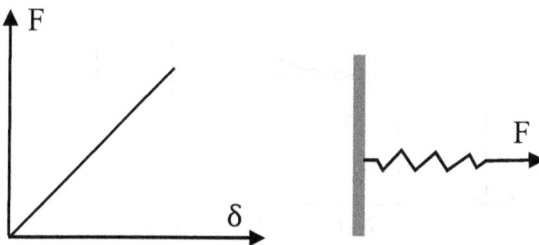

Fig. 11.1 Spring model.

In this case, the model has an explicit linear relationship between the force and the displacement. The displacement δ corresponding to a force F can also be calculated by the above formula if the stiffness k is known.

In many cases however, such relation is implicit or not well known at all. Direct calculation will be difficult or even impossible. In this situation, an approximate solution may be found numerically. Numerical simulation is an approach to search for an approximate solution based on certain correlations, as illustrated below.

Numerical approximation

In general, if a process involving one independent variable x and one dependent variable y, a relationship, either linear, non-linear relation or implicit, can be described by the following equation

$$y = f(x) \tag{11.2}$$

For multiple independent variables, the function $f(x)$ may contain two or more variables. If this relation represents the above spring model, the work done by the force from x = 0 to x = a is the area (A) below the curve as shown in Fig. 11.2 and can be determined by integration

$$A = \int_0^a f(x)\, dx \tag{11.3}$$

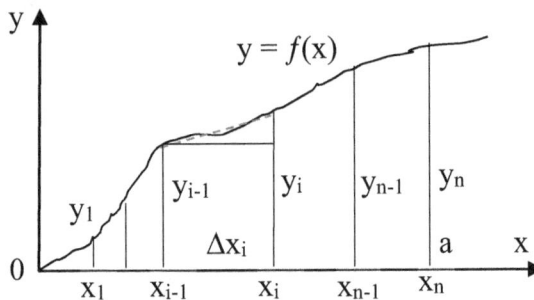

Fig. 11.2 Approximation of area calculation.

Integration can be performed only if the relation $f(x)$ is explicitly established. Otherwise, the area needs to be determined by other means. One approach is to estimate the area by numerical approximation.

First, let us divide the area under the curve $f(x)$ into N segments as shown in Fig. 11.2. The area is then estimated by summing up the areas of all segments

$$A \approx \sum_1^N A_i \tag{11.4}$$

where A_i is the area of the ith segment and can be calculated in several ways, such as

a) $A_i = y_{i-1} \Delta x_i$,
b) $A_i = y_i \Delta x_i$,
c) $A_i = (y_{i-1} + y_i) \Delta x_i / 2$,
d) $A_i = y_{i-1} \Delta x_i + (y_i - y_{i-1}) \Delta x_i / 2$.

In either way, there will be some difference between A_i and the actual value. In d), $f(x)$ is represented by a straight line and the triangular area is approximated by a right triangle. The result may be closer to the actual value, thus having the smallest error.

However, as the number of segments N increases, the difference between the estimated and actual values is reduced for all above methods. When N is large enough, the difference may eventually become negligible and the results become acceptable, no matter how A_i is calculated or whatever the relation $f(x)$ is in the ith segment. This approach is often called the finite difference method, which provides approximate but acceptable solutions.

At this point, we can summarize the steps we have followed to achieve an approximate solution:

Step 1 – Discretization. First, the area under the curve f(x) is divided into N segments. Note each segment does not necessarily have the same width. This is the process of discretization of a curve, a function or a domain to be analyzed. The purpose is to break a large problem into many smaller problems, which are then solved individually.

Step 2 – Approximation. Each segment is simplified based on the shape, and the relation of $f(x)$ is replaced by a straight line. An approximate solution is then calculated based on what we already knew by a properly simplified method. This is the process of approximation, which can be different in individual segments or regions.

Step 3 – Assembly and compatibility. To determine the final result, all solutions of individual segments are assembled with restrictions to ensure compatibility on the boundaries between segments. In this example, the same y_i value between adjacent segments is used.

Numerical modelling methods in geomechanics

Numerical modelling methods are based on the concept of approximation of a constitutional relation or a process to be analysed, but not necessarily a mathematical formula. In geomechanics, numerical modelling means the simulation of stress changes and deformation caused by excavations based on the constitutional relations governing the stress and strain / deformation in the rock mass. Equation 2.1 is the constitutional relation between stress and strain in one dimension. Equations 4.34 and 4.30 are for two and three dimensions, respectively.

A uniform stress field exists in the ground prior to excavation, which disturbs the balanced stresses due to removing a part of the rock. A process of stress redistribution occurs in the vicinity of the excavation to reach a new stress balance. Our interest is in how the stress is changed and how much is changed. In the field, this is a three-dimensional problem in general and may be simplified to a two-dimensional condition in a cross section of a long drift as shown in Fig. 11.3.

Let us consider a cross section of an underground drift, an example of a 2D condition. What will be the new stresses and deformation in the surrounding area after excavation?

We can follow the above steps to build a numerical model and find approximate solutions. The first step is discretization, which has very different approach in various numerical modelling methods. This is followed by finding approximation solutions in individual subregions and assembly of solutions to determine the final solution. However, the whole process for each modelling method is substantially different and can only be explained individually.

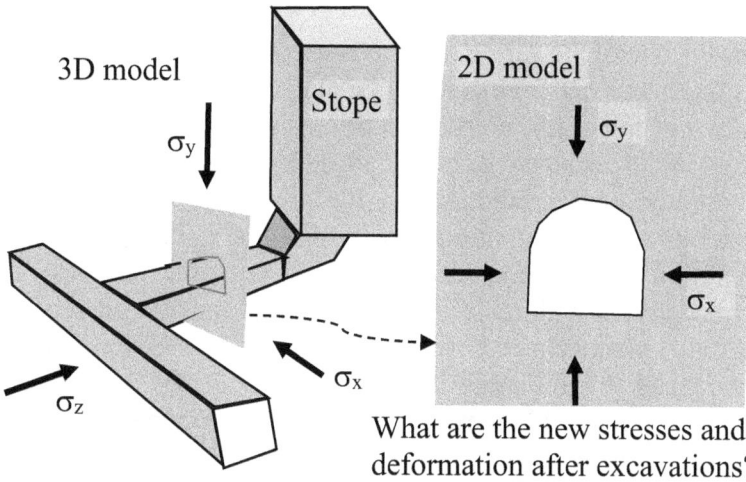

Fig. 11.3 3D model and 2D model of excavations

In geomechanics, the following numerical modelling methods have been commonly used: The Finite Difference Method (FDM, similar to the above example of numerical approximation), The Finite Element Method (FEM), The Boundary Element Method (BEM) and The Distinct Element Method (DEM). They may be used for 2D and 3D analysis.

In recent years, a Lateral-Spring Rock Mass model (LS-RM) is proposed for 3D modeling of rock mass, which is based on features of FEM and DEM and still under development. In geotechnical engineering, another Discrete Particle Method (DPM) has been used in recent years in 3D modelling, which is based on granular particles with bounding among particles, with similarities to DEM.

In the following sections, the basic principles for FEM, BEM and DEM in 2D modelling will be introduced, respectively and their applications in geomechanics discussed. For technical detail of each method and other methods, readers are suggested to read relevant textbooks and publications.

11. 2 Introduction to Finite Element Method

1) Basics of FEM

FEM is one of the earliest methods used in analysis of stress and displacement in engineering. Let us consider the 2D model shown in Fig. 11.3, a cross section of a drift. We are going to simulate the new stresses and displacements (including magnitude and orientation) in the area surrounding the drift under the given in-situ stresses. This is a 2D plane strain condition and consideration is given to stresses and displacements within the x-y plane without considering those off the x-y plane. In the following, a simplified approach is introduced to illustrate the concept of FEM. The actual formulation could however be more complex depending on the type of elements used (Cook 1974, Zienkiewicz 1977).

Model creation

To create a numerical model, a domain in a rectangular shape, large enough ($\geq$ 5 times of the drift size) to cover the entire affected area by excavation, is delineated and the drift is centered in the domain (Fig. 11.4). The drift is the inner

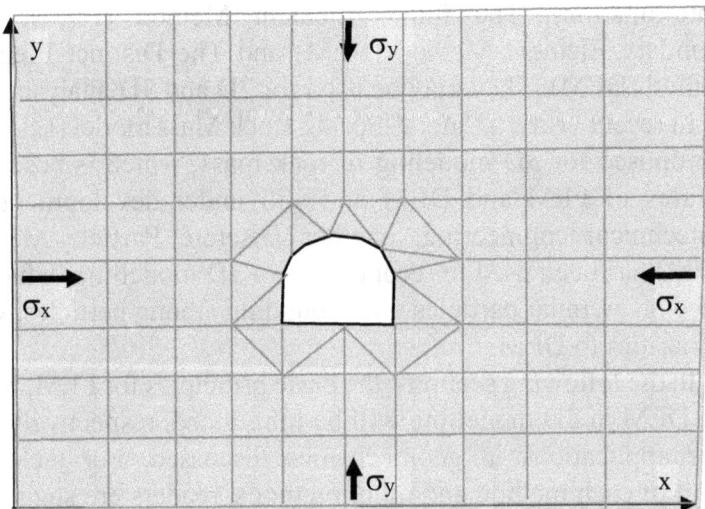

Fig. 11.4 Example of a 2D model mesh

boundary and the exterior edge of the domain is the outer boundary of the model. A cartesian coordinate system x, y is established as reference. In-situ stresses (σ_x, σ_y, τ_{xy}) are applied.

Discretization

The whole domain to be analyzed is divided into sub-regions, each being called an element. There are a total of M elements in the domain. Each element may be different in size and shape. In general, smaller elements are created at and near the excavation and larger elements at a farther distance. The result of discretization is a model mesh as shown in Fig. 11.4.

An element in the model mesh is defined by the coordinates at its corners (or vertices), which are called nodal points or simply nodes. A triangular element has three nodes and a rectangular element has four nodes (Fig. 11.5). There will be a total of N nodes in the model. A model may have a mix of different types of elements, even with non-straight edges. It is however important to have compatible edges between adjacent elements. Individual elements may have different properties and are treated individually.

Interaction among adjacent elements is only through the nodal points. All forces acting upon an element from, or transferred to adjacent elements are through the forces at its nodes (nodal forces). If gravity of an element itself is to be considered in analysis, it is assumed to act at the gravitational center of an element. For a medium with uniform mass, the geometric center is the same as the gravitational center.

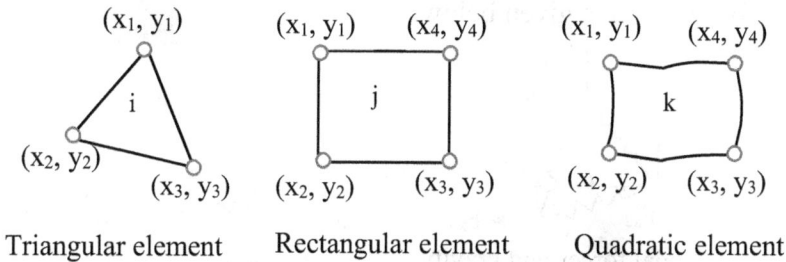

Triangular element   Rectangular element   Quadratic element

Fig. 11.5 Sample element shapes and nodal coordinates

Approximate solution

The solution to the stresses and displacement caused by excavation in the model is sought in two steps. The first step is to formulate the displacements at all nodal points. The second step is to calculate the displacements and stresses at various points within an element and the forces at nodes based on the results of nodal displacements, with consideration of the stress-strain relationships. However, there are no clear-cut boundaries between these steps in the analysis process.

In the first step of analysis, an element is represented by springs connecting each two of its nodes, as illustrated in Fig. 11.6. The rest of the element is viewed as "transparent", meaning not playing a role at this stage. An element in 2D is now represented by three or four 1D springs, which only transmit tensile or compressive forces along their axes, governed by Eqn. 11.1. Thus, the model mesh can be imaged as an assembly of many springs tied up at those nodes. External stresses acting upon the model and the unbalanced stresses caused by excavation are transmitted through these springs prior to reaching an equilibrium.

The primary variables in the analysis are the displacements and forces at the nodal points. In the x-y plane, they each have two components in the x and y directions, respectively. The nodal forces (F_x, F_y) and nodal displacements (u, v) are shown as positive in Fig. 11.6. They will have a negative value if in the opposite direction.

Once the whole model has reached a balance, the nodal forces and nodal displacements can be determined. More detail on formulation is given below.

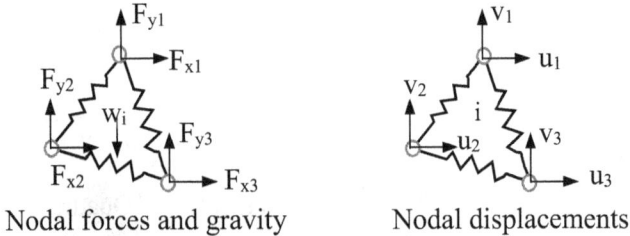

Nodal forces and gravity   Nodal displacements

Fig. 11.6 Spring model with nodal forces and displacements
(arrows indicate positive direction)

2) Formulation

Displacement formulation and nodal displacements

Displacement is considered here as a primary variable, upon which other parameters are determined. The actual displacement at a point in the domain is not explicitly known but it is expected to vary with its location (i.e., x, y coordinates). Let us assume the displacement component in the x direction at any point, u_x, as a polynomial function of its coordinates

$$u_x = a_1 + a_2x + a_3y + a_4x^2 + a_5y^2 + \ldots \qquad (11.5)$$

where x, y are the coordinates and a_i are called generalized displacement amplitudes, constants.

The above equation can written in a matrix form

$$u_x = [\phi]\,\{\alpha\} \qquad (11.6)$$

where $[\phi] = [1,\ x,\ y,\ \ldots]$ and $\{\alpha\} = [a_1,\ a_2,\ a_3, \ldots]^T$, the constant coefficients.

Similarly, the displacement component in the y direction at that point, u_y, can be given as

$$u_y = [\phi]\,\{\beta\} \qquad (11.7)$$

where $\{\beta\} = [b_1,\ b_2,\ b_3, \ldots]^T$, are also constant coefficients.

In the above equations, by choosing a higher order of x and y, more accurate solutions are expected. However, often the first order, or only linear terms are chosen as a simple approximation and the accuracy can be compensated by reducing the element size or increasing the number of elements.

If only linear terms (i.e., the first three terms) are chosen, let
$$\{A\} = [a_1,\ a_2,\ a_3, b_1,\ b_2,\ b_3]^T \qquad (11.8)$$
The displacement components can be written as

$$\begin{Bmatrix} u_x \\ u_y \end{Bmatrix} = \begin{Bmatrix} \phi_x \\ \phi_y \end{Bmatrix} \{A\} = [\Phi]\,\{A\} \qquad (11.9)$$

where $[\phi_x] = [1, x, y, 0, 0, 0]$ and $[\phi_y] = [0, 0, 0, 1, x, y]$.

Applying Eqn. 11.9 to the ith nodal point gives the two displacement components at that node $\{q_i\} = [u_i,\ v_i]^T$ as

$$\{q_i\} = \begin{Bmatrix} u_i \\ v_i \end{Bmatrix} = [\Phi_i]\ \{A\} \tag{11.10}$$

where $[\Phi_i]$ is determined with the nodal coordinates.

If Eqn. 11.10 is applied to all nodes, the nodal displacement vector $\{q\} = [u_1, v_1, u_2, v_2, \ldots, u_N, v_N]^T$, are given as

$$\{q\} = \underset{2N \times 1}{[\Phi_o]}\ \underset{2N \times 6}{\{A\}} \tag{11.11}$$

where $[\Phi_o]$ is an assembly of $[\Phi_i]$, $i = 1, \ldots N$ and is determined based on individual nodal coordinates. (The numbers below a vector or matrix represent its dimensions.)

Solving Eqn. 11.11 for the coefficients vector $\{A\}$

$$\{A\} = ([\Phi_o]^T [\Phi_o])^{-1} [\Phi_o]^T \{q\} = \{A_o\}\{q\} \tag{11.12}$$

where $\{A_o\}$, of size $6 \times 2N$, are determined by nodal coordinates.

Substituting Eqn. 11.12 into Eqn. 11.9 gives the displacement

$$\begin{Bmatrix} u_x \\ u_y \end{Bmatrix} = [\Phi]\ \{A_o\}\{q\} = [N]\{q\} \tag{11.13}$$

where $[N]$ is called the shape function and has a size of $2 \times 2N$.

Now the displacements at any point in the domain are given in terms of the nodal displacements vector $\{q\}$.

Stress-strain relations and stress formulation

In a plane strain condition, there are three strain components $\{\varepsilon\} = [\varepsilon_x, \varepsilon_y, \gamma_{xy}]^T$ corresponding to the three new stress components $\{\sigma^n\} = [\sigma_x^n, \sigma_y^n, \tau_{xy}^n]^T$. From Eqn. 4.34, they are correlated by

$$\{\sigma\} = [C]\ \{\varepsilon\} \tag{11.14}$$

where $[C]$ is the stress-strain relationship matrix and defined as

$$[C] = G \begin{bmatrix} k(1-\upsilon) & k\upsilon & 0 \\ k\upsilon & k(1-\upsilon) & 0 \\ 0 & 0 & 1 \end{bmatrix} \tag{11.15}$$

where $G = E/2(1+\upsilon)$ and $k = 2/(1-2\upsilon)$.

Strain is determined by differentiation of the displacement functions

$$\{\varepsilon\} = f(\dot{u}) = \left[\frac{\partial u}{\partial x}, \frac{\partial v}{\partial y}, \frac{\partial v}{\partial x} + \frac{\partial u}{\partial y}\right]^{T} \qquad (11.16)$$

Substitute Eqn. 11.13 into the above equation and simplify, we have

$$\{\varepsilon\} = [B] \{q\} \qquad (11.17)$$

where [B], size 3 x 2N, are determined by differentiation of the shape function matrix [N].
Substitute Eqns. 11.17 into 11.14

$$\{\sigma\} = [C] [B] \{q\} = [k] \{q\} \qquad (11.18)$$

where [k] has a size 3 x 2N.

Now the new stresses at any point in the domain are also given in terms of the nodal displacements vector $\{q\}$. If the nodal displacements are found, both the stresses and displacements in other locations can be determined.

Nodal forces and assembly of equations

In an element, considering stress balance, the stresses over the whole element can be represented by the nodal forces in x and y directions (F_x, F_y), as shown in Fig. 11.6. When the nodal displacements and nodal forces for all elements are assembled, Eqn. 11.18 eventually leads to

$$\{R\} = [K] \quad \{q\} \qquad (11.19)$$
$$\text{2N x 1} \quad \text{2N x 2N} \quad \text{2N x 1}$$

where [K] is called stiffness matrix and $\{R\}$ is a vector of all nodal forces in the same sequence as the nodal displacement vector $\{q\}$.

Based on the boundary conditions of the simulated domain, some of the nodal forces and displacements are known and the rest can be determined by solving Eqn. 11.19. Eventually all the nodal displacements can be determined. Finding solution to Eqn. 11.19 is one of the major tasks in FEM. Direct inversion of [K] may be a challenge because of the large matrix size and other methods such as Gaussian elimination may be used. Detail on how to solve large equations is beyond the scope of this textbook and users are referred to books on numerical solutions.

Calculation of displacements, stresses and strength factor

Once nodal displacements are determined, they can be used to calculate the displacements and stresses at any point by Eqns. 11.13 and 11.18, respectively. Usually, calculations are performed for a selected number of points within the simulated domain based on a grid.

The total displacement at a point can then be determined using vector addition

$$\mathbf{u_t} = \mathbf{u_x} + \mathbf{u_y} \tag{11.20}$$

The calculated new stresses (σ_x^n, σ_y^n, τ_{xy}^n) at a point are then converted to principal stresses (σ_1, σ_2) with their directions using Eqns. 4.10 and 4.8, respectively. Stress contours within the domain are then mapped according to stress magnitude.

Based on a selected failure criterion - Mohr Coulomb (c, ϕ) or Hoek's empirical method (m, s), a strength factor (or a safety factor) can be determined. A strength factor contour will indicate the stable and potential failure zones. More detail on these aspects will be demonstrated later in a section on applications.

11. 3 Introduction to Boundary Element Method

1) Basics of BEM

The concept of boundary element modelling is very different from that of finite element modelling. In BEM, only the excavation opening (the inner boundary) is discretized to N elements and the rest of the domain remains as a continuum as shown in Fig. 11.7. There is no outer boundary in a BEM model. However, a grid boundary is specified for calculation of stresses and displacements.

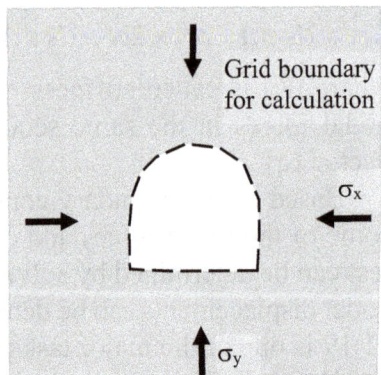

Fig. 11.7 BEM model and opening discretization

In a 2D BEM model, an element is a line segment along the excavation boundary. Excavation is realized by applying some sort of special condition on an element of the opening boundary, which is expected to affect the vicinity. The effects from all elements are superimposed on the initial stress field to arrive at a new stress condition due to excavation.

In BEM, there are three different approaches to realize excavation, giving rise to three sub-methods:

a) The fictitious stress method (stress is continuous on both sides of an element),

b) The displacement discontinuity method (different displacement may be applied to both sides of an element, useful for a thin crack where the first method fails),

c) The direct boundary integral method (based on the reciprocal theorem, or equal work).

Detail of BEM can be found in many textbooks (Crouch and Starfield 1983, Banerjee and Butterfield 1981). In the following, the principle of the fictitious stress method is discussed as an illustration of BEM.

Let's first look at the stresses on the excavation boundary prior to excavation. There are two pairs of stresses (σ, τ), one pair on each side of an element due to the in-situ stresses (σ_x, σ_y, τ_{xy}), and they are in balance as in Fig. 11.8a). If the rock mass to be excavated is isolated, its support to the surrounding rock mass can be viewed as those pairs of stresses (σ, τ) inside the boundary on all elements. In other words, the rock mass within the excavation boundary can be represented by those stresses inside the boundary. As such, excavation is virtually the removal of those pairs of inner stresses and can be achieved by applying negative stresses inside the boundary, which have exactly the same magnitude but in the opposite directions as illustrated in Fig. 11.8b).

Now suppose there are a pair of imaginary forces (F_{ni}, F_{ti}) acting on the ith element of the excavation boundary, $i = 1, .. N$. These are the fictitious forces we are looking for if they can cancel the original stresses inside the boundary. It should be kept in mind that any external force applied on an element will

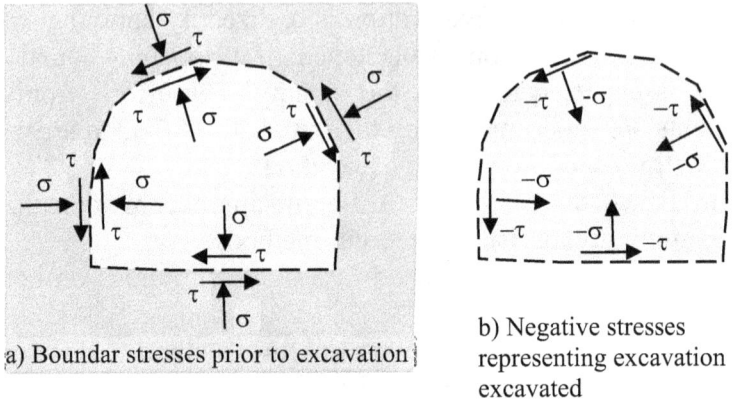

a) Boundar stresses prior to excavation

b) Negative stresses representing excavation excavated

Fig. 11.8 Stresses on excavation boundary

not only affect the element itself but also the other elements and the surrounding area. It is the combined effects of all fictitious forces together that will generate negative stresses of exactly the same magnitude to completely cancel the stresses inside the excavation boundary and represent excavation. Note the direction differences of (F_{ni}, F_{ti}) on each element: one perpendicular and one parallel to the element. They can only be added together after transformation to a common coordinate system, for example in the x and y directions. Thus, the new stresses and displacements surrounding an excavation are from the in-situ stresses plus the effects of the fictitious forces, as shown in Fig. 11.9.

2) Formulation

The problem of a concentrated force in an infinite elastic domain is typically a Kelvin's problem (Sokolmikoff 1956). In plane strain condition, assume a line of concentrated force is distributed along the z axis but it is a concentrated force in the x-y plane with two components (F_x, F_y), corresponding to x and y directions, respectively.

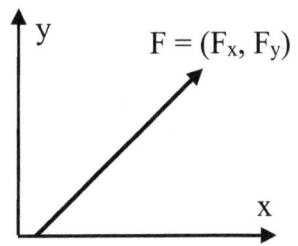

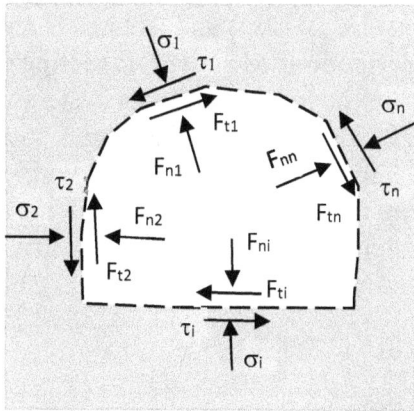

Fig. 11.9 Fictitious forces and stresses on excavation boundary

Omitting the details of deriving the solution to this problem, the final results are given in terms of the following function

$$g(x, y) = - k \ln \sqrt{x^2 + y^2} \qquad (11.21)$$

and its derivatives, where k is a constant.

If a concentrated force is evenly distributed along a line segment $|x| = a$, the solution is given in terms of an integration function

$$f(x, y) = \int_{-a}^{a} g(x - \xi, y) \, d\xi \qquad (11.22)$$

and its derivatives.

Based on the coordinates of a point, the above functions and their derivatives can be determined. The displacements (u_x, u_y) at a point, caused by the force are expressed as linear functions of F

$$\begin{Bmatrix} u_x \\ u_y \end{Bmatrix} = \begin{bmatrix} a_1 & a_2 \\ b_1 & b_2 \end{bmatrix} \begin{Bmatrix} F_x \\ F_y \end{Bmatrix} \qquad (11.23)$$

and the stresses induced by the force as

$$\begin{Bmatrix} \Delta\sigma_x \\ \Delta\sigma_y \\ \Delta\tau_{xy} \end{Bmatrix} = \begin{bmatrix} c_1 & c_2 \\ d_1 & d_2 \\ e_2 & e_2 \end{bmatrix} \begin{Bmatrix} F_x \\ F_y \end{Bmatrix} \qquad (11.24)$$

When the above solution is applied to an element on the excavation boundary in a local coordinate system with x' along and y' perpendicular to the element, the displacements u and v on the boundary (u // x', v $\perp$ x') and stresses σ and τ, ($\sigma \perp$ x', τ // x') caused by the fictitious forces (F_n, F_t) acting on the element can be determined in the local coordinate system with equations similar to Eqns. 11.23 and 11.24. When the fictitious forces on all of the N elements are considered, with proper stress transformation from ($\Delta\sigma_x$, $\Delta\sigma_y$, $\Delta\tau_{xy}$) to (σ, τ), the displacements and stresses at an element can be given in a matrix form in terms of N pairs of fictitious forces $\{F\} = [F_{n1}, F_{t1}, F_{n2}, F_{t2}, ,,, F_{nN}, F_{tN}]^T$

$$\begin{Bmatrix} u \\ v \end{Bmatrix} = [A_1]\{F\} \tag{11.25}$$

and $$\begin{Bmatrix} \sigma \\ \tau \end{Bmatrix} = [B_1]\{F\} \tag{11.26}$$

where $[A_1]$ and $[B_1]$ are of size 2 x 2N and $\{F\}$ is of size 2N x 1.

If Eqns. 11.25 and 11.26 are applied to every element on the excavation opening and equations are assembled together,

$$\{q\} = [A]\{F\} \tag{11.27}$$

and $$\{s\} = [B]\{F\} \tag{11.28}$$

where $\{q\} = [u_1, v_1 ... u_N, v_N]^T$, $\{s\} = [\sigma_1, \tau_1 ... \sigma_N, \tau_N]^T$. Both $\{q\}$ and $\{s\}$ have a size of 2N x1. [A] and [B] have a size of 2N x 2N.

On the opening boundary, some displacements and some stresses are known after excavation. For example, for unsupported excavation opening, $\sigma = 0$ and $u = 0$ on all elements. Eqns. 11.27 and 11.28 can be used to solve for the fictitious forces $\{F\}$.

Based on the direction of an element on the opening, the fictitious forces (F_n, F_t) on that element can be easily converted to (F_x, F_y). Then Eqns. 11.23 and 11.24 can be used to determine the displacements and stresses anywhere in the domain caused by the fictitious forces or excavation.

In linear elastic condition, the new stresses (σ_x^n, σ_y^n, τ_{xy}^n) can then be determined by superposition of the induced stresses ($\Delta\sigma_x$, $\Delta\sigma_y$, $\Delta\tau_{xy}$) and the in-situ stresses (σ_x, σ_y, τ_{xy}). For example, $\sigma_x^n = \sigma_x + \Delta\sigma_x$.

11. 4 Introduction to Distinct Element Method

1) Basics of DEM

Distinct element method is developed for analysis of jointed rock mass. While technical detail can be found elsewhere (e.g., Cundall 1971, Itasca 2014), some basics on DEM is presented below.

Creation of model mesh

Discretization in DEM is similar to that in FEM. A DEM model mesh also requires an outer boundary for the simulated domain and an inner boundary for the excavation opening. However, DEM allows natural joints to form solid elements (not wire mesh), called blocks as shown in Fig. 11.10. All blocks are numbered in a model.

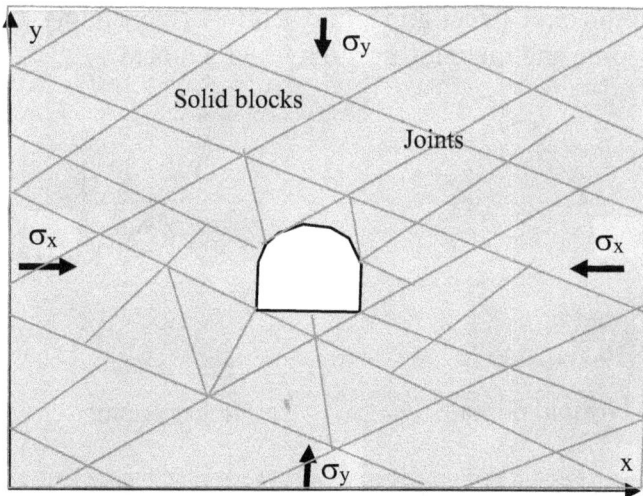

Fig. 11.10 Discretization and DEM model mesh in 2D

In a DEM model, any natural joints, boundaries between rock masses, bedding planes, etc. can be division lines between blocks. Solid blocks may be rigid (no deformation) or deformable. Each block may have its own properties (governing equation in the block and joint property at contacts with adjacent blocks).

Interface constitutive relations

Interaction between rigid blocks may be in one of the three forms: corner-corner, corner-edge and edge-edge and is represented by normal and shear contacts (Fig. 11.11), which are governed by the force-displacement law

$$\Delta F_n = K_n \, \Delta u_n \qquad\qquad (11.29a)$$

$$\Delta F_s = K_s \, \Delta u_s \qquad\qquad (11.29b)$$

where ΔF is the incremental contact force, Δu is incremental displacement at time increment Δt and subscripts n and s represent normal and shear contact, respectively.

The constants K_n and K_s represent the normal stiffness and shear stiffness, respectively. For boundaries between blocks in a rock mass and stronger joints, higher values are used. For weaker joints, lower values are used. $K_n = 0$ if separation occurs.

For deformable blocks, once in contact, there may be deformation and penetration into a block, depending on the contact force and the relevant strength of a block.

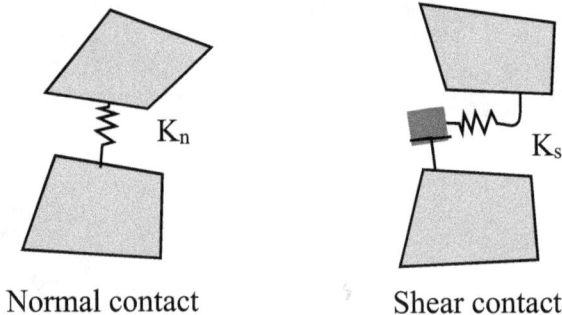

Normal contact   Shear contact

Fig. 11.11 Form of contact between rigid blocks

2) Solution procedure

DEM is very different in finding a solution from the previous two methods. In FEM and BEM, analysis was conducted under a static condition based on solid mechanics and the final solutions after excavation were found in one step. In DEM, analysis is conducted under a dynamic condition and the solution is arrived at following an iterative process based on a time increment Δt. Imagine a spring under an instant loading: it will bounce back and forth a number of times before it settles down. Displacement and stress vary during the process and eventually reach a balance.

Equation of motion

Solutions in DEM are explicit in the time domain and based on force balance on a block. Motion is governed by Newton's second law

$$\mathbf{F} = m\,\ddot{\mathbf{u}} \qquad (11.30)$$

where $\mathbf{F}$ is the total force vector acting upon a block, m is the block mass and $\ddot{\mathbf{u}}$ is the acceleration in the same direction as the force $\mathbf{F}$. If $\mathbf{F} \neq 0$, $\ddot{\mathbf{u}} \neq 0$. Then velocity $\dot{\mathbf{u}}$ and displacement $\mathbf{u}$ can be determined in time increment Δt.

Iteration procedure

In the explicit solution process, the first step is to determine the total force $\mathbf{F}$ acting upon a block from neighboring blocks. Then with consideration of gravity $\mathbf{g}$ and damping force if any, the acceleration is determined

$$\ddot{\mathbf{u}} = \mathbf{F}/m + \mathbf{g} - \alpha\,\dot{\mathbf{u}} \qquad (11.31)$$

where α is the damping coefficient. Once acceleration is known, it is applied to the centroid of the block (and also the grid points for a deformable block). The velocity $\dot{\mathbf{u}}$ and displacement $\mathbf{u}$ at the centroid and each corner of the block are calculated with a time increment Δt. Meanwhile, if the total force $\mathbf{F}$ does not act at the centroid of the block, possible rotation is determined at the corners of the block. The new coordinates of the centroid and all corners are then updated, and displacements are calculated. The new results of a block are applied to adjacent blocks and this process is conducted for all blocks in sequence. The strains, stresses and forces are updated based on the new displacements.

After completion of a round of iteration, if $\mathbf{F} \neq 0$, another time increment Δt is added, and the above process is repeated again and again until the whole system settles down. This is evaluated by checking the velocity $\dot{u}$ after each iteration. If $\dot{u} \approx$ 0 or $\leq$ a specified very small value, iteration stops. Otherwise, iteration continues. The displacements and stresses in the last iteration at the centroid and corners within a block are the final results. The displacements and stresses at other points within the block can also be calculated.

11. 5 Application of Numerical Modelling

In addition to the creation of a model mesh, conducting numerical modelling also requires information regarding field stresses and boundary conditions on the model. Computation and display / presentation of the modeling results (e.g., displacements, stresses, strength factor) are completed through computer software. However, interpretation of the results and assessment of ground stability are the users' responsibility. These aspects will be briefly discussed below and followed with demonstration using an example.

1) Field stresses

The stresses in the field are required for modelling and they need to be estimated first as discussed in Chapter 7. The stress field in-situ is in 3D, and proper stress components should be applied to a 2D model. When a model is created in a vertical plane, if at shallow depth, the field stresses to be applied to the model should include the gravity of the elements themselves plus the pressure from the overburden above the model. If at greater depth, the overburden pressure may be applied as constant field stresses at that depth. The gravity of the elements themselves may or may not be included depending on its influence in comparison with the overburden pressure.

When a model is created in a horizontal plane, the field stresses to be applied would be constant field stresses, which may be equal or greater than the overburden pressure at that elevation, without consideration of the gravity of the elements themselves. The gravity of the overburden is applied as an out of plane stress to the model if a modelling package is capable of doing so.

In 3D modelling, the complete 3D stresses corresponding to the model depth are to be applied to the model.

2) Boundary conditions

A model is a domain isolated from the "infinite" rock mass, and certain conditions need to be applied around the model to retain the model's status before and after isolation. These conditions are applied to the outer boundary for FEM and DEM models, but are not required for a BEM model because it has a different modeling principle.

The condition on the outer boundary varies with the model. It is supposed to reflect the original field condition but not be affected by excavation. This requires the model size to be large enough and at a sufficiently far distance from excavation. The distance from excavation to the outer boundary is usually 5 times or more of the size of the excavation opening.

For a model in a vertical plane, the rock mass on both sides of the model should be allowed to move up and down under gravity, but movement in the horizontal direction is restricted and this is achieved by using a roller boundary with $u_x = 0$. At the bottom of the model, movement may be restricted in both x and y directions using a fixed boundary (with $u_x = u_y = 0$), or only in the y direction using a roller boundary with $u_y = 0$. At the top of the model, if it coincides with the ground surface, it is left free. If it is below the ground surface, it is free to move in both x and y directions with no restrictions and a vertical stress at that elevation may be applied as well. A typical setup of the outer boundary condition is shown in Fig. 11.12.

If a model is created in a horizontal plane, the four sides may have the same conditions, roller or fixed.

The excavation opening is the inner boundary in a model. The condition of the opening after excavation needs to be considered as well. The inner boundary is usually left free after excavation, meaning no supporting pressure on the opening surface in the direction perpendicular to the boundary, $\sigma_n = 0$. This is true for an open stope. For a drift, even ground support is installed sometime after excavation, "free surface" may also be used. This is because when support is installed, deformation induced by excavation has already occurred, the support is supposed to resist further ground movement, but it can not restore the deformation which has occurred.

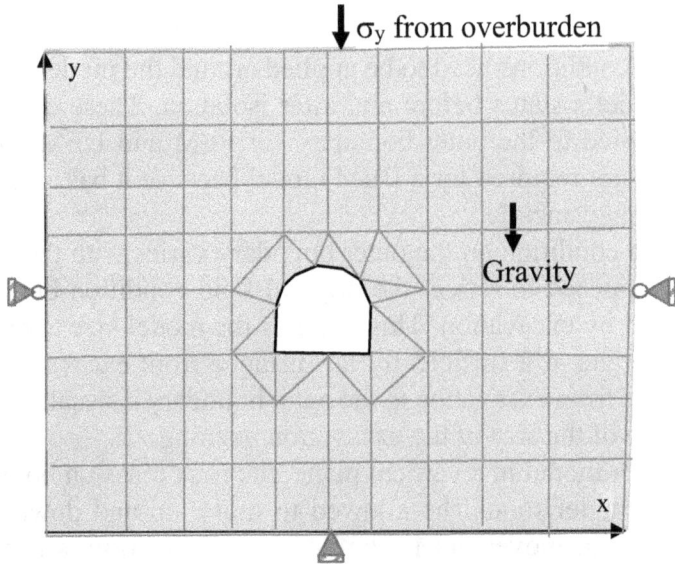

Fig. 11.12 Boundary conditions on the outer boundary in a vertical plane

3) Modelling software

As shown above, numerical modelling involves complex formulation and requires tremendous amount of computation. Creation of a model mesh also requires the delineation of individual elements and definition of all nodes. These tasks can only be accomplished with specifically designed computer software. There are numerous commercially available modelling software packages, each developed based on a specific modelling method. Some may be designed for specific applications and others may be for multiple purposes.

A numerical modelling software package is usually a stand-alone application software without use of other software (e.g., Excel or Matlab). It is normally written in a computer language, such as C++ (or Fortran in the early days). A modelling software package usually includes four modules: user interface, mesh generator, computation and data interpretation, although users can only interact with the first and the last modules and have no access to the other two.

A user interface is the first part of modelling software. It allows users to define problems to be solved, enter geometry, rock properties, boundary conditions, field stresses and other required parameter for modelling. What a user can see is only this, and even data interpretation is often imbedded in the interface as one of its functions.

A mesh generator generates a model mesh, defines elements and calculates nodal coordinates automatically. Computation (solving all relevant equations to determine the needed primary parameters) is done automatically once data input is complete. This is the core of a modelling software package but users are not able to see it. Users may be required to click on a function button, like "compute", to complete the task.

How good a package is and how reliable the results are can only be assessed from the produced results. Generally, a commercial package is tested vigorously to ensure its functionality and accuracy. However modelling results can only be as good as the knowledge of the user and the input data.

4) Presentation of modelling results

Results presentation is the last and a very important part of a modelling software package. It is usually part of the interface as a built-in function and users will not see any additional execution file. However, this module will function only after computation is successfully completed. Its purpose is to show users the modelling results, normally in graphics (for example, contour plots of new stresses, displacements, safety factor, etc.) and allow users to export and/or print the results.

The direct parameters in modelling results are usually the new principal stresses (σ_1, σ_2) and the total displacement u_t (or sometimes displacement components u_x and u_y) at select points within the domain. The number of points where results are calculated depends on the grid size, which are also automatically generated to give an appropriate presentation of the results.

These results are normally plotted as smooth contour lines so that users will not see "gaps between points".

Although the main goal of modelling is to simulate new stress distributions and displacement induced by excavation, the ultimate goal is to assess the ground stability around the excavation. This is achieved by comparing the new stresses with the rock mass strength at a location based on one of the failure criteria discussed in earlier chapters to determine a safety factor (also called strength factor). Rock mass strength may be determined by the Mohr - Coulomb criterion using (c, ϕ) parameters or the Hoek's empirical criterion using (m, s) constants. A safety factor less than one means failure. Therefore, a contour plot of safety factors can delineate potentially unstable and stable zones and help users assess ground stability. However, one must remember that the final responsibility rests on the users, not on the modelling results. If a user cannot interpret the modelling results appropriately, his/her endeavor to modelling is not a success.

5) Example of numerical modelling application

In the following, an example is introduced to illustrate how numerical modelling is applied for stress analysis in underground mining. Figure 11.13 is a model with dimensions and excavation location. This model simulates an underground mine. A stope was mined and backfilled in the upper level and a new stope at 1000 m depth is designed underneath a sill pillar. A service drift exists at the sill pillar elevation and a new haulage drift is designed in the footwall at the lower level. Stress changes and ground stability in the vicinity need to be assessed.

A model in the vertical plane was created with a model size at least 10 times of the excavation size to encompass the whole excavation area. Roller boundary was specified at the bottom and on both sides. The top was left free to move. A field stress of $\sigma_h = 1.5\sigma_v$ was applied, with σ_v being the overburden pressure at 1000 m depth.

Modelling was conducted using FEM simulation. The model mesh is generated using a software package. The input data included model size, opening size and geometry, field stresses and rock mass properties.

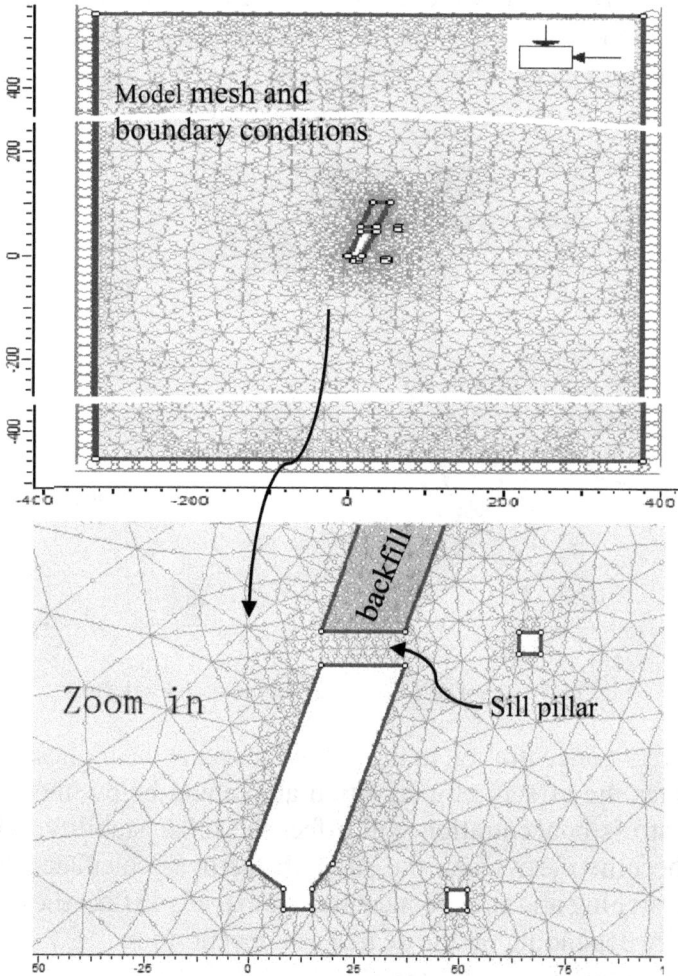

Fig. 11.13 Sample model mesh

Sample modelling results in the area surrounding the excavations are displayed in Figs. 11.14 to 11.17. The major principal stress σ_1 is shown in Fig. 11.14. It confirms high stress concentration around the corners and in the sill pillar after excavation. The contour lines indicate the magnitude of σ_1. The stress trajectory shows the overall change of the new principal stress directions (longer arms for σ_1 and short arms for σ_2). Prior to excavation, σ_1 is horizontal and σ_2 is vertical in the whole domain. After excavation, σ_1 becomes nearly vertical and parallel to the stope forming a stress arch on both sides of the stope as well as on the top and bottom.

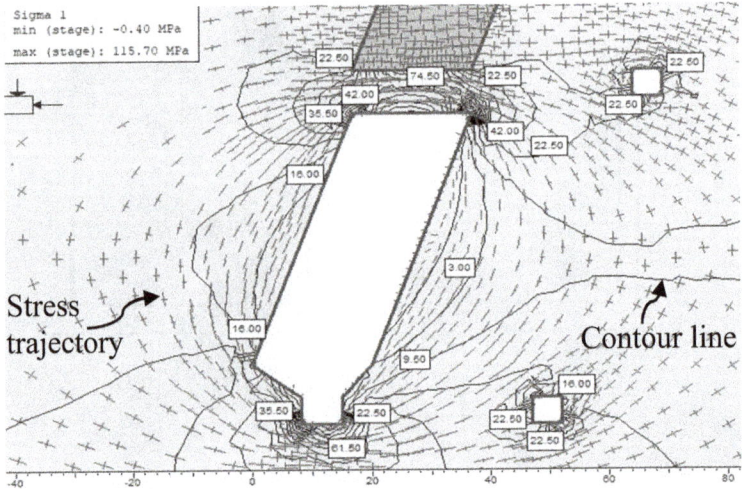

Fig. 11.14 Major principal stress contour and trajectory

Stress concentration in the sill pillar seems the highest and it indicates the effects of excavation above and below the pillar. High stress concentration will affect the pillar stability, which will become clear when the strength factor is calculated later. Stress arching and stress concentration may also affect the drifts' stability depending on the degree of concentration.

The minor principal stress in Fig. 11.15 shows tension ($\sigma_3 <$ 0) in the hangwall and footwall in the upper portion of the stope, confirming what is expected. In those areas, tensile failure may exist, causing ore dilution.

Figure 11.16 is the enlarged total displacements, which all point towards the open stope. The sill pillar tends to move down and may collapse if it becomes unstable.

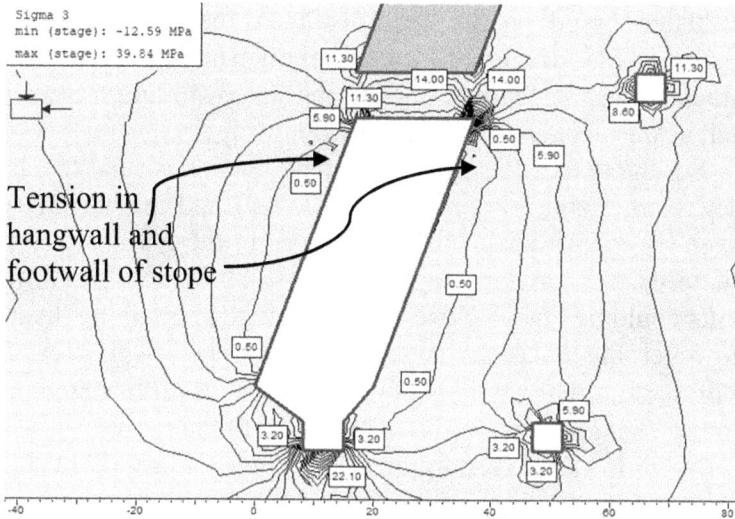

Fig. 11.15 Minor principal stress contour

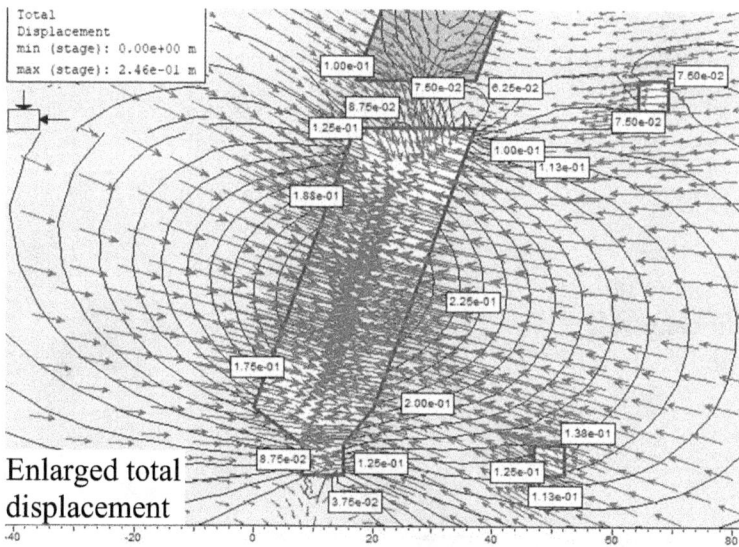

Fig. 11.16 Total displacement contour and direction

The overall stability surrounding the excavation area is evaluated based on the specified rock mass properties. The results of the strength factor are shown in Fig. 11.17, which indicated tensile failure zones in the hangwall and footwall, as well as unstable areas at the draw point.

For the two drifts, other than small tension around the drifts, they seem stable (strength factor > 1.0) and are not affected much by excavation of the two stopes in the given stress field where $\sigma_h > \sigma_v$. After examining Figs. 11.14 and 11.17, these drifts could be placed closer to the stope to shorten the distance. However, the situation may be different if the in-situ stress field is of $\sigma_h \leq \sigma_v$. The effect can be evaluated by further modelling, simply changing the field stress applied to the model.

Note: In the above figures, the notes and arrows were not generated by the software but added afterwards by the author as part of results interpretation. It is up to you, the user, to interpret what those stress and displacement results mean. As mentioned earlier, the results can only be as meaningful as the knowledge of the user.

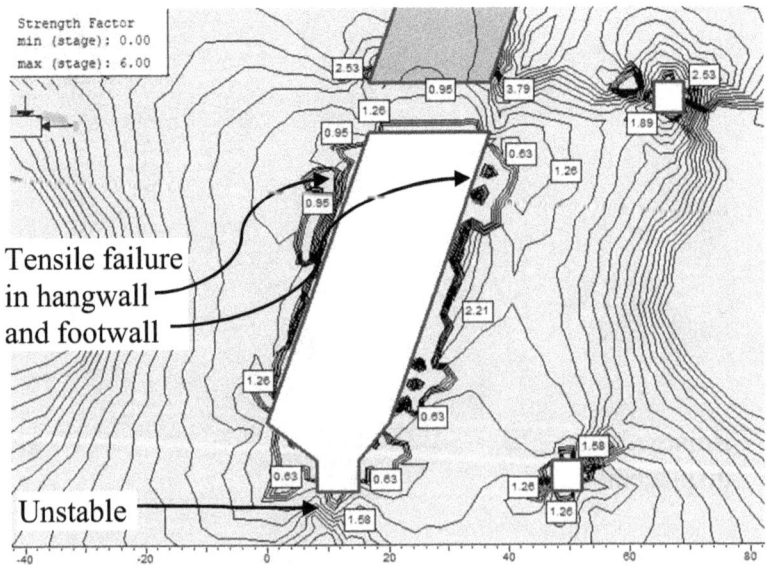

Fig. 11.17 Strength factor contour

5) Discussions

Several aspects need to be noted:

a). As the above results indicate, at sharp and acute corners of an excavation, high stress concentration occurs. This is not the expected nor the desired outcome of excavation. In actual practice, such sharp corners should be avoided, or may not actually exist considering how blasting and machine excavation are completed. In modelling, a sharp corner can be avoided by replacing the corner node with two nodes and a small element, or by an arch element when the model is created.

b). Numerical modelling itself is an approximation approach. The results will have some difference from the unknown "true" values. In addition, the input data contain uncertainties: ground condition is not very well known, rock mass properties and field stresses are estimated, etc. As a result, the modelling results cannot and should not be expected to be "perfect" or precisely "accurate". Some software may allow users to pick up specific points (e.g., around an opening) and give "exact" values of stress and displacement generated by modelling. One should exercise extreme caution when it comes to evaluating the results at a specific point / location, particularly on the excavation boundary.

This is however not to say that numerical modelling results are not useful. It is useful if the results are interpreted knowledgeably. For a complex situation as in the above example, it would be very difficult for us to imagine the stress changes around the stopes and the effect on the drifts. Modelling results give us a clear picture on what is happening in the area and can indicate the trend of change.

In multiple stages of excavation or mining sequencing, numerical modelling is perhaps the most useful tool to help us visualize the trend of stress changes and the progress of ground stability as excavation and mining advance.

References

Banerjee, P.K. and R. Butterfield, 1981. Boundary element methods in engineering science. McGraw-Hill, London, 452p.

Barton, N., R. Lien and J. Lunde, 1974. Engineering classification of rock masses for the design of tunnel support. *Rock Mechanics*, **6(4)**: 189-236.

Barton, N., 1976. Recent experiences with the Q-system of tunnel support design. In *Proc. Symp. on Exploration for Rock Engineering*, Johannesburg, **1**: 107-117.

Barton, N.R., 1987. Predicting the behaviour of underground openings in rock. Manuel Rocha Memorial Lecture, Lisbon. Oslo: Norwegian Geotech. Inst.

Bieniawski, Z.T., 1976. Rock mass classification in rock engineering, In *Proc., Symp. on Exploration for Rock Engineering*, Johannesburg, **1**: 97-106.

Bieniawski, Z.T., 1989. Engineering rock mass classifications. New York, Wiley.

Brady, B.H.G. and E.T. Brown, 2006. Rock Mechanics. Springer, Netherlands, 628p.

Byerlee, J.D., 1968. Brittle ductile transition in rocks. *J. of Geophysics*, **73(14)**: 4741-4750.

Cook, R.D., 1974. Concepts and applications of finite element analysis. Wiley, NY, 402p.

Crouch, S.L. and A.M. Starfield, 1983. Boundary element methods in solid mechanics. Allen & Unwin, London, 322p

Cundall, P.A., 1971. A computer model for simulating progressive large-scale movements in blocky rock systems. In *Proc. Symp. Int'l Society for Rock Mechanics*, **1**: II-8.

Deere, D.V., 1964. Technical description of rock cores for engineer purposes. *Rock Mechanics and Engineering Geology*, **1(1)**: 17-22.

Golder Associates, 1981. Prediction of stable excavation spans for mining depths below 1000 m in hard rock. Report to CANMET, DSS Serial No. 05480-0081.

Griffith, A.A., 1924. Theory of rupture. In *Proc. 1ˢᵗ Congress Appl. Mech.*, Delft, 55-63.

Griggs, D.T., 1936. Deformation of rocks under high confining pressures: I - experiments at room temperature. *J. of Geology*, **44(5)**: 541-577.

Griggs, D.T., 1939. Creep of rock. *J. of Geology*, **47**: 225-251.

Griggs, D.T., F.J. Turner and H.C. Heard, 1960. Deformation of rocks at 500° to 800° C. *Rock Deformation*, **79**: 39-104.

Grimstad, E. and N. Barton, 1993. Updating of the Q-system for NMT. In *Proc. Int'l Symp. on Sprayed Concrete*, Fagernes, 46-66.

Grimstad, E., K. Kankes, R. Bhasin, A. Magnussen and A. Kaynia, 2002. Rock mass quality Q used in designing reinforced ribs of sprayed concrete and energy absorption. In *Proc. Int'l Symp. on Sprayed Concrete*, Davos, 134-142.

Heidbach, O., M. Tingay, A. Barth, J. Reinecker, D. Kurfeß, and B. Müller, 2008. The World Stress Map database. http:/www.world-stress-map.org.

Heidbach, O., M. Rajabi, K. Reiter, M. Ziegler, 2016. World Stress Map (2016). GFZ Data Services. http://dataservices. gfz-potsdam.de/wsm/showshort.php?id=escidoc:1680899.

Herget, G., 1988. Stresses in Rock. Rotterdam, Balkema.

Hoek, E. and E.T. Brown, 1980a. Underground Excavations in Rock. Institution of Mining & Metallurgy, London, 527p.

Hoek, E. and E.T. Brown, 1980b. Empirical strength criterion for rock masses. *J. Geotech. Eng. Div.*, ASCE **106(GT9)**: 1013-1035.

Hoek, E. and E.T. Brown, 1988. The Hoek-Brown failure criterion – a 1988 update. In *Proc. 15th Canadian Rock Mech. Symp.*, 31-38.

Hoek, E., P.K. Kaiser and W.E. Bawden, 1995. Support of underground excavations in hard rock. A.A. Balkeama, Brookfield, 215p.

Hoek, E., D. Wood and S. Shah, 1992. A modified Hoek-Brown criterion for jointed rock masses. In *Proc. Rock Characterization Symp. Int Soc. Rock Mech.*, 209-214.

ISRM (Int'l Society for Rock Mechanics), 1981. Rock characterization, testing and monitoring – ISRM suggested methods. Oxford, Pergamon.

ISRM, 1989. Suggested method for determining point load strength, RTH 325-89: 55-60.

Itasca Consulting Group, 2014. UDEC user manual. https://www.itascacg.com/software/udec.

Jaeger, J.C. and N.G.W. Cook, 1976. Fundamentals of rock mechanics. Chaman and Hall, London, 585p.

Jeremic, M.L., 1994. Rock mechanics in salt mining. A.A. Balkema, Rotterdam, 532p.

Kompen, R., 1989. Wet process steel fibre reinforced shotcrete for rock support and fire protection. Norwegian Practice and experience. In *Proc. Underground city*, Munich, 228-237.

Lauffer, H. 1958. Gebirgsklassifizierung fur den Stollenbau. *Geologie und Bauwesen*, **24(1)**: 46-51.

Lin, C., 2019. Estimation of 2D and 3D in-situ stresses using back analysis of measurements of well/borehole deformation. Ph.D. thesis, Dalhousie University, Canada, 213p.

Loset, F., 1997. Engineering geology – practical use of the Q-system. NDI report 592046-4.

Merritt, A.H., 1972. Geologic prediction for underground excavations. In *Proc. 1st N. American Rapid Excavation and Tunnelling Conf.*, AIME, New York, 601-622.

NGI (Norwegian Geotechnical Institute), 2013. Using the Q-system - handbook. http://www.ngi.no/en/Contentboxes-and-structures/Reference-Projects/Reference-projects/Q-system/.

Patton, F.D., 1966. Multiple modes of shear failure in rock. In *Proc. 1st congr. Int. Soc. Rock Mech.*, Lisbon, **1**:509-513.

Potvin, Y., M. Hudyma, and H.D.S. Miller, 1988. The stability graph method of open stope design. Presented at *the 90th CIM AM*, Edmonton.

Serafim, J.L. and J.P. Periera, 1983. Consideration of the geomechanical classification of Bieniawski. In *Proc. Int. Symp. on Engineering Geology and Underground Construction*, Lisbon, **1(II)**:33-44.

Singh, B., J.L. Jethwa and A.K. Dube, 1992. Correlation between observed support pressure and rock mass quality. *Tunnelling & Underground Space Technology*, **7(1)**: 59-74.

Singh, M. and B. Singh, 2008. High lateral strain ratio in jointed rock masses. *Engineering Geology*, **98**(3-4): 75 - 85.

Sokolmikoff, I.S., 1956. Mathematical theory of elasticity. McGraw-Hill, NY, 476p.

Song, Z., W.M. Xiao, H.Y. Ni and G. Fan, 2015. Large lateral deformation characteristics of simulated columnar jointed rock mass under uniaxial compression tests. *Int'l J. Geohazards & Environment.* **1(3)**: 122-127.

Stillborg, B., 1994. Professional users handbook for rock bolting (2nd Ed.). Clausthal-Zellerfeld: Trans Tech Publications.

Terzaghi, K., 1946. Rock defects and loads on tunnel supports. *Rock Tunnelling with Steel Supports*, Ed. R. Proctor and T. White, Commercial Shearing and Stamping Co., 15-99.

Trueman, R., P. Mikula, C. Mawdsley and N. Harries, 2000. Experience in Australia with the application of the Mathews method of open stope design. *CIM Bull.*, **93(1036)**: 162-167.

Vreede, F.A., 1981. Critical study of the method of calculating virgin rock stresses from measurement results of the CSIR triaxial strain cell. Research report, Pretoria, South Africa.

Windsor, C.R., 1992. Cable bolting for underground and surface excavations. In *Proc. Int'l. Symp. on Rock Support*, Sudbury, Balkema, 349-376.

Zienkiewicz, 1977. The finite element method (3rd Ed.). McGraw-Hill, London, 787p.

Zou, D.H. and P.K. Kaiser, 1990. Determination of in situ stresses from excavation-induced stress changes. *Rock mechanics and Rock Engineering*, **23**: 167-184.

Zou, D.H., 2004. Rock bolt loading mechanism. In *Proc. SINOROCK 2004 - Int'l Symp. on Rock Mechanics*, ShanXia, China, (**3A 08**) on CD-ROM.

Appendices

Rock Mechanics
Laboratory Manual

I.   Rock Sample Preparation

II. Tensile Strength by Indirect Brazilian Test

III. Point Load Strength Index Test

IV. Shear Strength on Rock Surfaces by Direct Shear Test

V. Uniaxial Compressive Strength, Elastic Modulus and Poisson's Ratio by Unconfined Compression Test

VI. Triaxial Compressive Strength and Internal Friction Angle by Confined Compression Test

Appendix I

Rock Sample Preparation

1. Objectives

a) to be familiar with the procedure and equipment for rock sample preparation,
b) to learn the skills for preparing rock cores.

2. Tasks

Prepare rock samples for specific type of tests. Rock sample preparation usually involves coring, cutting, and polishing when necessary. The requirements of dimensions and number of samples for various tests are listed in Table A.1.

Table A.1 Requirements of rock specimens.

Test	Dimensions	Number*
Uniaxial compression	Diameter d, NX core (~2", 54mm) Length L = 2.5~3 d (4.5" ~ 5")	≥ 3
Triaxial compression	NX core, Length L = 2~2.5 d (4" ~ 4.25")	≥ 3
Tensile strength by Brazilian test	NX cores, Disk thickness t ≅ 0.5 d	≥ 3
Point load strength	Core sample: Diametrical loading: L ≥ 1.0 d Axial loading: 0.3 < L/ d <1.0	≥ 10 cores
	Lump sample (Fig. A.7) :	≥20 lumps
Shear strength on rock surfaces	Core samples can be used. Length: to fit specific test devices (see shear section).	2~3 pairs

* The indicated numbers of specimens are required for standard tests and for each type of rock. For demonstration purpose, they can be reduced.

Once rock samples are cut to a specific length, further preparation may be needed to make specimens for a specific test. More detail on specimen preparation for each test will be given in the subsequent sections of this manual.

If rock cores are available from diamond drilling in the field, such as those shown in Fig. A.1, they may be used if they meet the above requirements. In this case, cores can be cut to the required length.

Fig. A.1 Rock cores from field drilling.

Rock samples should be free from obvious defects, such as joints, bedding planes, damage from chipping or crushing, etc. Collection of rock samples, either cores or blocks, to be used for tests, should be representative of the rocks in the field.

To conduct laboratory tests, students are divided into groups, with 3 to 5 students in each group. Every group should prepare their own rock samples and specimens, do tests and submit own reports.

Laboratory Safety

Laboratory safety cannot be overemphasized. Throughout laboratory sessions, heavy equipment and rocks are involved. All students must be equipped with safety gear and follow instructions of technicians and teaching assistants.

3. Rock sample preparation

Coring procedure

If no core samples are available, samples can be made from rock blocks on a coring machine as shown in Figure A.2, where a core sample is already made.

Fig. A.2 Rock coring machine in the Dalhousie University Rock Mechanics Laboratory.

In general, the rock block to be used for making core samples should be in a rectangular prism or cubic shape, at least 4.5" thick to meet the requirements of Table A.1. The path on the rock block where the core barrel is going through should be free of defects.

To drill cores, a rock block is placed on the bottom chaise of the coring machine. The chaise can be moved up and down, and be swung left and right to align the coring barrel at a proper location on the rock. The rock block is secured in place with metal bars before coring. It is recommended to place a piece of plywood at the bottom of the rock to avoid damaging the drill bit when drilling is through the rock.

Before the start of coring, turn on water switch slightly to deliver water to lubricate the coring bit. At the beginning of drilling, guide the coring barrel by holding the bottom of the barrel with one hand while it rotates at low speed. Always wear protective gloves. The coring bit will make an annular hole in the rock. Once the annular hole is deep enough to hold the barrel in position, the coring machine can be set to automatic advancing until coring is complete.

When coring is complete, retreat the coring barrel up. A gentle tap on the barrel will release the cored sample inside the barrel. Use a hand at the bottom of the barrel to catch the sample to avoid damaging the sample from free fall.

Cutting rock samples to the proper length

Depending on the type of test, a rock sample is to be cut to a specific length as required in Table A.1. This is accomplished with a diamond saw as shown in Fig. A.3.

A rock sample is first cut at one end to remove the uneven end surface. It is then measured and marked at a proper length. The rock sample is secured in place by a pair of clamps, which can be loosened or tightened by a threaded bolt. The saw blade is aligned with the marked position and the clamps are then tightened.

If a rock sample is too short to hold in the clamps properly, place another short piece of core, in line with the sample to be cut, between the clamps to help secure the sample in place.

The cutting blade is lubricated with oil. As shown in Fig. A.3, a portion of the blade is submerged in an oil tank at the bottom. This is supposed to cool the blade as it cuts the rock. The cover of the machine must be properly closed before turning on the power switch.

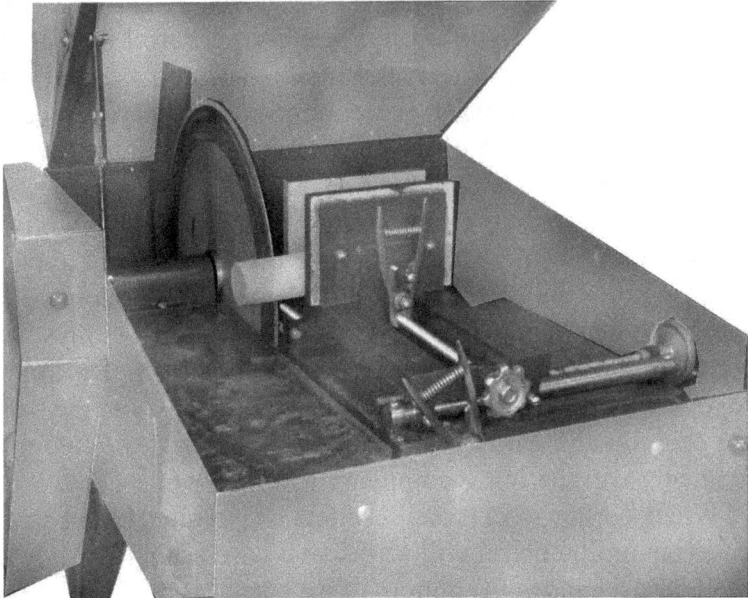

Fig. A.3 Rock cutting machine in the Dalhousie University Rock Mechanics Laboratory. A sample is ready to be cut.

Polishing end surfaces

Once rock samples are cut to a specific length, they need to be polished on both ends for uniaxial and triaxial compression tests. (Polishing is usually not necessary for other tests). The saw-cut ends of a sample should be parallel to each other and perpendicular to the sample axis, with a tolerance of deviation less than $0.25°$. This is achieved by polishing both ends on a polishing machine, as shown in Fig. A.4.

Polishing is done by a grinding disk, which is mounted on the machine but adjustable up and down. A rock holder, which can hold up to four cores through bolts, keeps the cores perfectly perpendicular to the moving table underneath the edge of the grinding disk. The rock holder is secured on the moving table by an electric-activated magnet. The table is able to move left and right, and in and out, to align the rock samples with the grinding disk as polishing progresses. Table movement can be operated manually by a skilled person, or automatically by the machine.

Before polishing, both ends of a core sample should be marked with lines using a marker. Polishing is complete when all lines have disappeared.

The polishing machine is lubricated with water or specific fluid. It must be turned on to lubricate the grinding disk before polishing starts.

Fig. A.4 Rock polishing machine in the Dalhousie University Rock Mechanics Laboratory. A sample is ready to be polished.

Specimen storage

Rock samples collected in the field should be stored, normally not more than 30 days, in such a way to preserve the natural water content until specimen preparation time. After preparation, specimens should be stored in room temperature (20°±2°C) with 50%±5% humidity for 5 to 6 days before testing.

Appendix II

Tensile Strength by Indirect Brazilian Test

1. Objectives

a) to become familiar with the Brazilian test procedure,
b) to learn the method of determining tensile strength by compressive loading.

2. Tasks

Prepare test specimens to meet the specifications for the Brazilian test and conduct tests on specimens to determine tensile strength.

3. Specimen preparation

NX cores are preferred. Cut cores to make disk specimens to meet the specification of the thickness/diameter ratio, t/d = 0.5. Both ends of a specimen should generally be smooth, flat, parallel to each other and perpendicular to the specimen axis within 0.25°. The sides should be smooth and free of abrupt irregularities.

Measure specimen diameter in three different directions around and calculate the average diameter to the nearest 0.1 mm. Follow the same step to determine the average specimen length to the nearest 1.0 mm.

4. Apparatus

The apparatus required to conduct the Brazilian test consists of two steel loading jaws, designed so as to contact a disk-shaped rock specimen at diametrically-opposite surfaces over an arc of contact of approximately 10° at the point of failure. An illustration is shown in Fig. A.5. The apparatus basically has three pieces: the upper and lower steel jaws as well as a small half ball bearing. The jaws are in an arch shape with a much larger diameter than the specimen and are aligned with bolts on both sides. This allows the jaws to move freely in the vertical direction. A disk specimen sits between the upper and lower jaws with "a line contact" with the jaws. The lower jaw is flat at the bottom. The upper jaw has a small spherically shaped dent on the top, where a half ball bearing will sit. This will allow the loading forces on the upper and lower jaws to align.

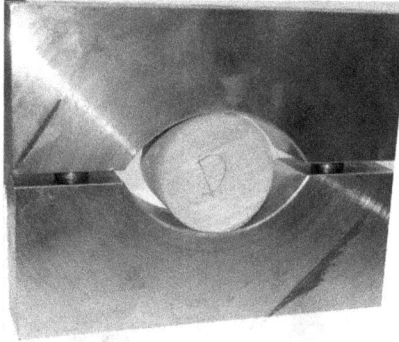

Fig. A.5 Brazilian test apparatus.

During the test, the apparatus together with the specimen is placed on a compressive loading frame, such as the one shown in Fig. A.6. The loading frame includes the upper loading head which is fixed to the frame but able to turn around spherically, the lower platform which can travel vertically under hydraulic pressure, a pumping system and load recorder. The pumping system activates the lower platform to move upwards towards the upper fixed head, creating compressive stress on the rock specimens in between. If the hydraulic pressure is released, the lower platform can travel downwards.

5. Testing procedure

Wrap the prepared disk specimen around its periphery with one layer of masking tape, which will help prevent local crushing at the contacts between the specimen and the loading jaws.

Place the specimen between the loading jaws, as shown in Fig. A.5, with the specimen aligned in the proper direction. The vertical plane through the specimen center is where tensile failure is expected to take place and it should be free of defects.

Fig. A.6 Compressive loading frame in the Dalhousie University Rock
Mechanics Laboratory.

Assemble the apparatus together and place it with the specimen on the lower platform of the compressive loading frame, with proper vertical alignment. Start the loading frame manually to raise the platform to a position where the upper loading head is just about to touch the half ball bearing on the apparatus. Then set the loading rate at a constant value such that the weakest rocks will fail within 15 to 30 seconds. A rate of 200 N/s is recommended.

If devices are available, the displacement and lateral expansion may also be recorded by installing devices before loading starts. Apply load to the specimen until it fails. The loading force at failure should be recorded.

Note: the result is acceptable only if the specimen is split through the central vertical plane at failure. Otherwise, the result is not valid.

6. Calculations and reporting

The tensile strength is calculated by Eqn. 3.7 (see Chapter 3)

$$\sigma_t = \frac{2P}{\pi dt} \tag{3.7}$$

where P is given in unit N, d and t in meter, and the result of strength is usually converted to MPa. The result to be reported should be the average of all specimens tested from the same type of rock.

When preparing a report, the following information should be included:

- lithology of the rock sample,
- presence of any weakness, for example, joints, bedding planes, foliation, etc.,
- orientation of the loading direction with respect to the specimen anisotrophy,
- dimensions and number of specimens tested,
- water content and storage history,
- test duration and loading rate,
- time and date of test,
- description of the failure mode on the failure surface,
- tensile strength of each specimen and the average.

Example

A number of disk samples are tested under Brazilian testing conditions. The sample size and the load at failure are listed in the table below (left four columns). Determine the average Tensile Strength from the data.

Specimen	d [in.]	t [in.]	P [lb]	σ_t [MPa]
U-1	2.01	0.495	3350	14.80
U-2	2.018	0.495	3203	14.09
U-3	2.01	0.49	3530	15.75
U-4	2.012	0.5	3671	16.04
U-5	2.016	0.5	3277	14.29
				15.00

Solutions:

Calculate the tensile strength for each sample (note unit conversion), as follows:

By Eqn. 3.7,

$$\sigma_t = \frac{2P * 9.8/2.2}{\pi d * 0.0254 * t * 0.0254}$$

The results are given in MPA, shown in above table (right column). The average tensile strength is 15 MPa.

Appendix III

Point Load Strength Index Test

1. Objectives

a) To learn the test procedure for point load strength,
b) To determine Point Load Strength Index $I_{s(50)}$ on rock specimens.

2. Tasks

Prepare test specimens to meet the specifications described in Table A.1, conduct tests on specimens to determine point load strength, and calibrate the results against the specimen size to determine $I_{s(50)}$ at a standard size of 50 mm.

3. Specimen preparation

Specimens for point load tests can be rock cores or lumps, as shown in Fig. A.7. A minimum of 10 specimens are required for

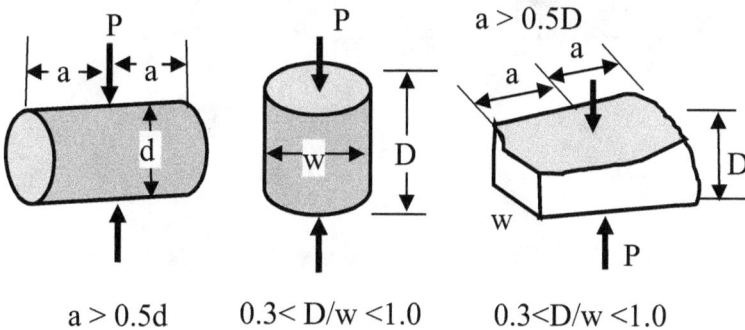

Fig. A.7 Point load directions and measurements.

cores in each test direction and 20 for rock lumps. Cores are the preferred specimens as they are simply cut to the required length.

If no cores are available, rock lumps are collected in the field. Lumps should be approximately rectangular blocks, satisfying the requirements. If lump samples are used, a portable test machine may be taken to the field for on-site testing.

4. Apparatus

Figure A.8 shows a portable point load tester. It can be easily taken to the field. It basically consists of a manually-operated hydraulic pump, a pair of spherically-truncated conical platens (upper platen is fixed to the frame and lower platen is movable under hydraulic pressure), a load recorder and a ruler to measure the distance between the two loading contacts.

Fig. A.8 Point Load Tester in the Dalhousie University Rock Mechanics Laboratory.

5. Test procedure

Record specimen dimensions and sort the specimens into groups based on the loading direction. Core specimens with a length/diameter ratio greater than 1.0 are suitable for diametrical testing, and otherwise for axial testing.

Insert the specimen between the two loading platens, carefully raise the lower platen by hand pump to make contact with the specimen. For diametrical testing, make sure the distance from the contact point to either end is at least 0.5d. For axial testing or lump testing, make sure the contact points are centered on the specimen.

At this point, put the safety guard (a wire mesh box, for example) in place to cover the specimen and device to prevent injury from flying rock at the time of specimen failure.

Increase load by gradual and steady pumping until the specimen fails. Record the failure load. The result is acceptable if the specimen broke through its mass. If failure occurred by chipping off small pieces in local areas, the data is invalid.

6. Calculations and reporting

Calculate the point load strength index I_s by Eqn. 3.16

$$I_s = P / D_e^2 \qquad (3.16)$$

where P is the load at failure, D_e is the equivalent diameter of the cross sectional area through platen contact points. For diametrical test, $D_e = d$. For other tests, D_e is determined by Eqn. 3.17.

$$D_e^2 = 4 \, w \, D / \pi \qquad (3.17)$$

Report the I_s result in MPa. Then use Eqn. 3.18 to correct the size effect for the calibrated result at a standard size of 50 mm:

$$I_{s(50)} = I_s \, (D_e / 50)^{0.45} \qquad (3.18)$$

where D_e is in mm.

When preparing a report, it is suggested to list all data and results in a table. The following information should be included:

- specimen identification,
- lithology of the rock sample,
- presence of any weakness and orientation of the loading direction with respect to the weakness,
- specimen dimensions,
- water content and storage history,
- time and date of test,
- description of the failure mode,
- calculated result of I_s and $I_{s(50)}$ and the average result,
- any other information as necessary.

The following is an example of a reporting table.

Point Load Test Report
_____
Date: Tested by:
Description of samples and other necessary information here.

No. Rock type	w	D	P	D_e	I_s	$I_{s(50)}$
a). Diametrical test						
a1 sandstone	-	2"	20kN	2"	?	?
a2 sandstone						
Average $I_{s(50)}$:		Standard deviation:				
b). Axial loading						
b1 sandstone	2"	1.8"	17kN	?	?	?
b2 sandstone						
Average $I_{s(50)}$:		Standard deviation:				
c). Lump test						
c1 sandstone	2.5"	2.1"	22kN	?	?	?
c2 sandstone						
Average $I_{s(50)}$:		Standard deviation:				

Other information:
_____

Example.

Point load tests are conducted on several core samples under diametrical loading. The sample size and failure load are shown in the table below (left three columns).
   Determine the average Point Load Strength Index (Hint: determine the individual I_s first, convert to $I_{s(50)}$ and then the average).

Specimen	d [in.]	P [lb]	D_e^2 [m²]	I_s [MPa]	$I_{s(50)}$ [MPa]
1-B	1.68	4820	1.821E-03	11.79	10.98
2-B	1.74	5760	1.953E-03	13.14	12.43
3-B	1.7	4190	1.865E-03	10.01	9.37
4-B	1.7	6360	1.865E-03	15.19	14.22
5-B	1.82	5300	2.137E-03	11.05	10.66

Average: 11.53

Solutions.

   Under diametrical loading, $D_e = d$. Calculate D_e^2, I_s and $I_{s(50)}$, (note unit conversion), as follows:

$$D_e^2 = (d*0.254)^2 \ \ [M^2],$$

by Eqn. 3.16, $I_s = (P*9.8/2.2)/ D_e^2 \ \ [MPa]$,

by Eqn. 3.18, $I_{s(50)} = I_s \ (d*25.4/50)^{0.45} \ \ [MPa]$.

   The results are shown in above table (right three columns). Then calculate the average of $I_{s(50)}$, which is 11.53 MPa.

Appendix IV

Shear Strength on Rock Surfaces by Direct Shear Test

1. Objectives

a) to learn the procedure for direct shear test on rock surfaces,
b) to determine the peak and residual shear strengths on saw-cut rock surfaces

2. Tasks

Prepare test specimens for direct shear test, conduct shear tests in forward and reversed directions at different normal stress levels, and analyze testing data to determine shear strength parameters.

3. Specimen preparation

Measure the depth of the lower half of the shear specimen mould (Fig. A.9, right), the half with a metal bar at the bottom.

Fig. A.9 Shear specimen mould.

Take a core sample, make a mark across the cutting location, and preferably cut the core in half at an inclined angle to the core axis with a cut surface in an elongated shape. The cut surface should have an area of approximately 6 in². (A core cut perpendicularly to the axis may be accepted for demonstration purpose). The length of each half core, measured perpendicularly to the cut-surface, should be slightly shorter (~ 0.25") than the depth of the mould, in order to fit in it. Any excess length should be cut off.

Record the core information of rock type, presence of joints, direction of core axis and cut direction with respect to them.
Put the two half cores back together at their original position with the saw-cut surfaces in good contact, wrap them with steel wires in 2 directions across the core and tighten with pliers.

Insert the core in the clamps (Fig. A.9, on the right) and hand tighten the clamps on both ends to hold the core in position.

Place the lower half of the shear mould on a level table, replace and tighten the two end plates by hand on both sides of the mould. Apply a thin layer of grease to the interior surface of the mould and the plates to allow easy removal of cement later.

Sit the clamps with core together on the flat edges of the shear mould and adjust the position of the core so that the saw-cut surface is parallel to but slightly higher than the edges of the shear mould. That is the plane where shearing will take place during the test.

Using a container, add a pre-calculated amount of water, cement and aggregates, and mix them thoroughly. Or simply add water to the pre-mixed concrete mix. The mix should be slightly runny for easy handling.

At the lower sides of the lower half shear mould, there are two threaded bolts. Turn the bolts in a few mm so that the bolts will be embedded in the cement. Use a trowel or a scoop to fill the shear mould, where the specimen is already, with the mixed cement to flush with the edges of the mould.

Check the specimen again to make sure the saw-cut surfaces are properly positioned. Leave the prepared half shear specimen in the mould for at least 24 hours to allow cement to cure.

Reposition the upper half of the shear mould on top of the lower half, align and tighten all bolts to hold them together. Turn the assembled shear mould upside down carefully. The two bolts should hold the cemented specimen in place to prevent falling.

Fill the other half of the shear box with the fresh cement mix again. Let it cure for at least 24 hours.

Loosen the nuts on the plastic plates on both sides of the shear mould. Remove the cement-specimen block from the mould carefully. For the lower portion of the shear mould, unscrew the two bolts first before attempting to remove the cemented specimen. The two halves of the cement-specimen blocks should be one assembly with the two halves tied together. Store it until testing time.

4. Apparatus

The shear test device used in this test consists of a shear box, a test frame, two hand pumps and measuring gauges. The shear box has two halves (exactly the same size and shape as the shear mould) to hold the two halves of the prepared cement-specimen block. The upper half shear box is shown in Fig. A.10 and the lower half is in the test frame (Fig. A.11). The test frame is hydraulically activated and is equipped with three pistons, one at the top and two on both sides at the bottom.

Fig. A.10 Upper half of the shear box and half shear specimen.

Fig. A.11 Direct shear test device in the Dalhousie University Rock Mechanics Laboratory, with shear box in place (pumps not shown).

During the tests, the lower half of the shear box is fixed to the test frame and stays still. The upper half sits (upside down) on top of the lower half and is attached to one of the two pistons on the frame. The attached piston will pull the upper half shear box to one direction. After completion of shearing in one direction, the piston is disengaged and another piston is connected to pull the upper half shear box to the opposite direction in turn.

One pump supplies pressure to the top piston to apply a normal pressure on the shear specimen surface. Another pump supplies pressure to the other two pistons (one at a time) to apply shear stress on the specimen surface. Both pumps are manually operated.

There are three measurement gages: one dial gauge to read the shear movement in micrometers, and two pressure gauges connected to the two pumps to indicate the pressures. A fourth dial gauge may also be used to measure the dilation (movement perpendicular to the shear surfaces) when necessary.

5. Testing procedure

With the two halves of the cemented shear specimen tied together, put it inside in the lower half of the shear box on the test frame, with the lower half at the bottom. Properly align them to ensure the saw-cut surfaces of the specimen are parallel to the shear box edges in the horizontal plane. The cemented specimen should sit in the shear box tightly. If there is difficulty in inserting the cemented specimen into the shear box, remove some cement on both ends using a sharp knife or chisel.

Replace the upper half of the shear box on top of the cemented specimen and align them properly. Assemble other components together as well.

Connect the pump that supplies normal pressure to the test frame. Then apply some pressure gradually to the pump to allow the top piston to touch the shear box and exert a small amount of pressure on it to prevent it from moving.

Use a hack saw blade, a cutter or a similar device, through the gap between the upper and the lower shear box, to cut off the steel wires which were used to hold the two half specimens together. Because of the limited access and lighting, this is a challenging task and requires patience.

Connect another pump to the test frame and make sure all switches are in the proper position. Apply a little pressure to allow the piston to contact the pull chain of the upper shear box properly.

Position all gauges and reset to zero if possible. Take initial readings of all gauges.

Supply pressure using a hand pump to increase the normal pressure to a pre-determined level and keep it constant.

Supply shear pressure manually and gradually, at a very small increment. Stop at each increment to allow reading of pressure and shear displacement simultaneously. This needs cooperation of three people: one to apply as well as read shear pressure, one to read shear displacement and one to record the data.

Increase shear pressure again to the next increment and take readings again. Repeat the process.

Stop loading when the upper shear box starts slipping. Slip is initiated when you can no longer increase the shear pressure to a higher level and the pressure keeps dropping back each time you try to increase pressure.

Be very alert to capture the initial maximum shear pressure because it disappears very quickly from the gauge. That will be the pressure level for the peak shear strength. The pressure that remains low when slip has been initiated will be the pressure level for the residual shear strength.

With everything in place as is, turn the necessary hydraulic switches to allow the upper half of the shear box to move in a reversed direction. You may reposition the displacement gauge or use a separate gauge.

Repeat the test to finish reversed shearing.

Use a new specimen, set normal pressure at another level and repeat the tests. The test should be conducted on at least three specimens at three different normal stress levels.

6. Calculations and reporting

The pressure gauges will give the direct readings of force or stress. Plot all data, from the forward and reverse tests, to show shear stress versus shear displacement. An example of shear stress-shear displacement curves is shown in Fig. A.12.

The first highest point on the curve corresponds to the peak shear strength. The curve fluctuates after that point. The average level corresponds to the residual shear strength.

There will be some difference in the forward and reverse shear test data. Use both curves to determine the average:

$$\tau_s = (\tau_{s1} + \tau_{s2})/2 \qquad\qquad \text{(A.1a)}$$

$$\tau_r = (\tau_{r1} + \tau_{r2})/2 \qquad\qquad \text{(A.1b)}$$

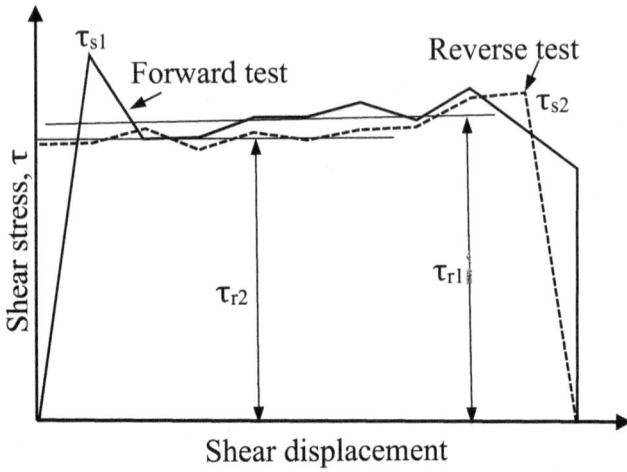

Fig. A.12 Typical shear stress – shear movement plot.

Plot the results of peak and residual shear strength against normal stress from all tests. An example is shown in Fig. A.13. Draw a best-fit line through those points, one for the peak strength and one for the residual strength data. The residual strength line should start from the origin and the peak strength line will intersect the τ axis at c. The slope angle of each line will corresponding frictional angle. These two best-fit lines are given by the following equations:

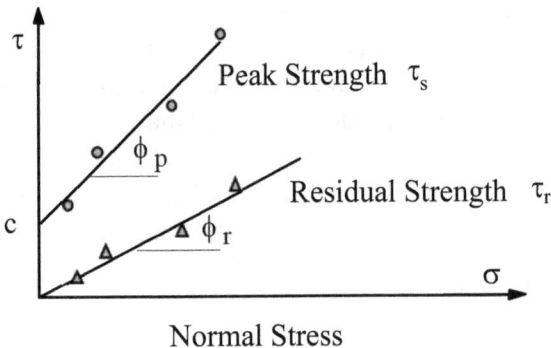

Fig. A.13 Typical shear strength versus normal stress from shear test.

$$\tau_s = c + \sigma \tan \phi_p \tag{3.9}$$

$$\tau_r = \sigma \tan \phi_r \tag{3.10}$$

The shear strength coefficients (c, ϕ_p, ϕ_r) are then determined.

Note: Unless the τ - σ chart is plotted on exactly the same scale on both axes, the slope angle cannot be measured directly on the chart and must be calculated based the coordinates.

When preparing a report, it is suggested to list all data and results in a table and give the analysis of data to show the results and charts as shown above. The following information should also be included:

- specimen identification,
- lithology of the rock sample,
- presence of any weakness and orientation of the loading direction with respect to the weakness,
- specimen dimensions,
- water content and storage history,
- time and date of test,
- average values of (c, ϕ),
- any other information as necessary.

Appendix V

Uniaxial Compressive Strength, Elastic Modulus and Poisson's Ratio by Unconfined Compression Test

1. Objectives

a) to learn the procedure of the uniaxial compression test,
b) to determine uniaxial compressive strength σ_c, Young's modulus E and Poisson's ratio υ.

2. Tasks

Prepare rock specimens for uniaxial compression tests, attach strain gauges to specimens, conduct tests and analyze the data to determine σ_c, E and υ.

3. Specimen preparation

Cut NX cores to meet the specifications given in Table A.1. Finish the end surfaces on the polishing machine. The side surface of specimens should be dry with no defects.

Attach strain gauges to specimens:

Strain gauges are attached to the specimen side surface in pairs. Two pairs are recommended for each specimen. For each pair, one gauge is parallel and another perpendicular to the specimen axis, as shown in the margin diagram.

Find a proper location on the side surface of the specimen where no visible fracture or defects exist. Follow the instructions in the strain gauge package to clean the

rock surface with the specified cleaning detergent. Use a white cloth to wipe and clean the surface. When the cloth shows no stains and the surface is dry, the surface is ready to attach strain gauges.

Take two strain gauges and measure their resistance with a meter to make sure they are not damaged. Strain gauges have one side coated. Make sure to attach the uncoated side to the specimen surface. Strain gauges are very small, approximately 10 mm long × 5 mm wide and very thin (like a plastic sheet). Handle them carefully with the proper tools.

Apply bond and attach strain gauges very carefully as per strain gauge instructions. This is a very delicate work. Extreme caution and patience must be exercised during the process. As the initial setting time of the bond is less than 60 seconds with a complete curing time of a few minutes, you need to get everything in place and be ready to complete the task promptly. Observe carefully when a technician or a teaching assistant is giving a demonstration.

Use scotch tape to cover the strain gauges that are attached to the specimen surface, leaving the two lead points exposed.

Clean the lead points with the specified cleaning fluid. Do not touch them with your hands or any dirty objects. Cut two pieces of wire, approximately 4' – 5' long. Strip the plastic off the ends of the wires (approximately 5 mm long), apply rosin to both the lead points and the wire ends. Solder carefully the wires to the lead points separately. There is a very small gap between the two lead points. Do not solder them together, this will short circuit the gauge. A specimen with two strain gauges attached is shown in Fig. A.14.

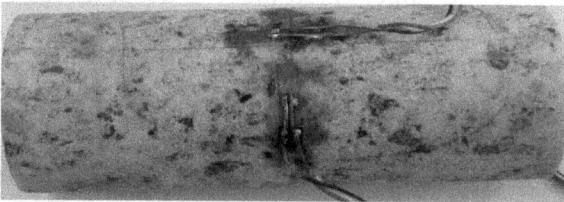

Fig. A.14 Example of rock specimens with strain gauges attached.

Use scotch tape to secure the wires to the specimen, to help avoid damage. Test the wires with a meter to make sure there is proper connection with the gauges. There should be a small resistance reading. If it is zero, it is shorted. If the resistance is too large, the wire or the gauge may be broken, or the soldering may not be done properly. In either case, re-do the work to correct the problem.

4. Apparatus

For unconfined compression testing, no special apparatus is needed other than the loading frame. A specimen can be placed directly on the loading platform as shown in Fig. A.6. To ensure the load is applied parallel to the specimen axis, the upper fixed loading head has a spherical ball bearing which allows the head to move and shift freely around a central point.

However, for data recording, a sophisticated device may be required. The reading device could be a completely automated data logging system to record load and all strains simultaneously and continuously during loading. In this case, computer software is required to process data and plot the results.

A manually operated device may also be used. In this case, two people are needed to work together, one person operating the pump to the loading system and another person taking readings of stress and strains at each step of loading.

5. Testing procedure

Measure the specimen dimensions: length and diameter, in at least three places and calculate the averages.

Place the specimen with strain gauges attached between the lower platform and upper fixed loading head, as shown in Fig. A.6 and align the specimen centrally on the platform.

Turn on the pump and fast raise the lower platform to allow the specimen to almost touch the upper fixed head. Then gradually apply a seating load, equivalent to approximately 1% of the estimated compressive strength, to allow the whole assembly to sit properly. Stop the pump.

Connect wires of the strain gauges to the data logging system, or to a strain-indicator device, whatever the case may be. Remember the channels connected to the axial and circumferential gauges. Set all strain readings to zero. If a zero setting is not possible, take the initial readings and subtract them from each of the future readings to determine the actual strains.

Close the metal screen gate (or any other safety guards) on the loading frame to prevent injury from flying rocks.

Turn on the loading pump and set a proper loading rate to start compression loading.

From this point onward, if an automated data logging system is used, the pump is left on to allow the compression load to increase at the pre-set rate until the specimen fails. Make sure the data logging system is recording load and strains simultaneously. If a manual recording device is used, increase the compression load in small increments (e.g., at 5% of the estimated strength) and stop the advancing switch (or the pump) temporarily to allow recording of load and strain. Then restart the switch/pump to continue loading by another increment. Record data again. Repeat this process until the specimen fails.

The loading rate should be set properly. The duration of approximately 5 to 10 minutes for the whole loading period (until failure) is considered appropriate in continuous loading. Alternatively, a loading rate between 0.5 and 1.0 MPa/s is acceptable.

6. Calculations and reporting

Transfer all data to an excel sheet, with time, load and strain sequentially in separate columns.

The data of the time series are based on the sampling rate of the automated data logging system. They are only for reference and will not be used in calculations. For manually recorded data, this may not be available.

Convert the data series of load force to stress, by dividing the load P by the cross section area of the specimen

$$\sigma = P / (\pi\, d^2 / 4) \tag{A.2}$$

The strain data are normally recorded in micro-strain, i.e., $10^{-6}\varepsilon$, although it has no unit. Check strain gauge instructions to confirm.

Plot the data as $\sigma \sim \varepsilon$ for the axial and circumferential strains in separate curves but on the same chart, as illustrated in Fig. A.15. The two strain series would have opposite signs, normally the axial strain positive and the other negative. You may convert them all to positive values for better appearance on the chart, as shown in Fig. A.15.

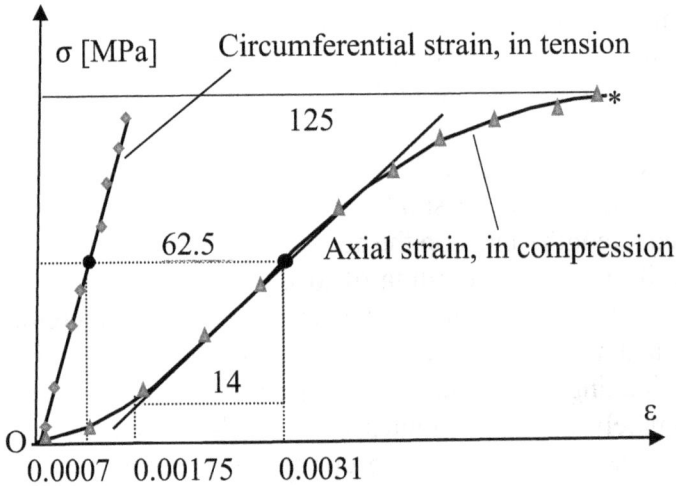

Fig. A.15 Example of stress-strain curves.

Follow the Example 3.1 in Chapter 3 to determine the uniaxial compressive strength σ_c, the tangential modulus E_t, secant modulus E_s, and Poisson's ratio υ.

Calculate average values from the above results for all specimens and report them in a table. The report should also include a description of the failure mode of each specimen, with sketches or photos, as shown in the margin. Compare the angle of the failure plane with that predicted from a Mohr's failure envelope discussed in Section 5.3.

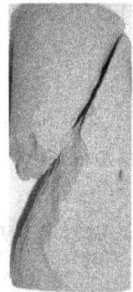

Appendix VI

Triaxial Compressive Strength and Internal Friction Angle by Confined Compression Test

1. Objectives

a). to learn the procedure of the triaxial compression test,
b). to determine the triaxial compressive strength and internal friction angle of rock specimens.

2. Tasks

Prepare rock specimens for triaxial compression tests, conduct tests at different confining pressures and analyze the data to determine the triaxial strength and the internal cohesion c_i and friction angle ϕ_i.

3. Specimen preparation

The test specimens are prepared using NX cores based on the specifications given in Table A.1. Cores are cut to the proper length and both ends are polished as described in Appendix I. Specimens should be dry and free from defects and damage.

A good practice is to conduct three tests on nearly identical specimens at three different confining pressures, with three specimens in each test. The bare minimum is three specimens from each rock sample which are tested at three different confining pressures. The alternative is to have nine tests on nine specimens from the same rock sample at nine different confining pressures covering the stress range to be investigated.

The exact diameter and length of specimens will depend on the triaxial compression cell to be used (see below). Ideally the specimen should be easily inserted into the triaxial cell but not fit too loosely. It should be slightly shorter than the flexible membrane in the cell. Specimens of 2.125" diameter (or NX core) and 4" to 4.25" length are appropriate.

4. Apparatus

For triaxial testing of rock specimens, the equipment will include a loading frame (Fig. A.6) and a triaxial compression cell, as shown in Fig. A.16, where a rock specimen and two steel platens are also shown in sequence related to their positions.
The loading frame should have sufficient capacity to apply enough loading force that was estimated, based on the specimen diameter and rock strength.

Fig. A.16 Triaxial compression cell in the Dalhousie University Rock Mechanics Laboratory, showing a rock specimen and two steel platens.

The triaxial compression cell is normally designed for the standard specimen size (e.g., NX cores). Figure A.16 shows the triaxial cell used in the Dalhousie University Rock Mechanics Laboratory. Inside the cell is a cylindrical impermeable flexible membrane, which is sealed at both ends. The annulus space between the triaxial cell body and the membrane is an oil chamber.

During tests, the triaxial cell is connected to a hydraulic system with a hand pump, and the rock specimen is completely inserted in the membrane with one steel platen at each end. The platen at the bottom has flat ends and the platen on the top has two sections which sit together on spherical ends to allow proper alignment of loading force. When the oil chamber is filled with pressured oil, a confining pressure is applied through the membrane on the side surface of the specimen. A pressure gauge connected to the hydraulic system will indicate the confining pressure.

5. Testing procedure

Measure the specimen dimensions: length and diameter in at least three places and calculate the averages.

Place the triaxial compression cell flat on a table. Insert the specimen in the cell at the center. Insert the steel platens at both ends in contact with the specimen (see Fig. A.16). Ideally, a small section of the platen should be inside the membrane.

Connect a hydraulic hose to the cell and apply a small amount of pressure manually to the cell to hold the specimen in place (it is tight enough if the specimen inside the cell cannot move).

Measure the total length of the assembled triaxial cell with the specimen and two steel platens in place. Measure the gap between the lower loading platform and the upper fixed loading head on the loading frame (Fig. A.6). Add or remove steel discs/spacers to adjust the gap to a height slightly greater than the length of the assembled triaxial cell.

Turn the assembled triaxial cell in the vertical position and carefully place it on the loading frame (replacing the rock specimen shown in Fig. A.6) between the platform and the loading head. Since the triaxial cell assembly is heavy with many attachments, this work may require assistance of another person.

Align the assembly centrally on the loading platform and apply a small axial load to the assembly, approximately 1% of the estimated strength. Check to make sure that the assembled triaxial cell is centrally and vertically aligned on the platform.

If necessary, attach a displacement measurement device in parallel to the assembled triaxial cell to measure the axial deformation.

Close the safety gate. Apply confining pressure to the triaxial cell at the pre-determined level. At the same time, an axial load approximately matching the confining pressure is applied simultaneously. Record the axial load at this point as P_o and axial displacement as zero.

With the confining pressure (σ_3) being constant, start increasing axial load gradually at a proper constant rate until specimen failure. A loading rate between 0.14 and 1.0 MPa/s, or a loading duration for failure to occur within 5 to 10 minutes of continuous loading, is acceptable.

If a stress-strain curve is required, both axial load and axial displacement should be recorded simultaneously during the testing. Axial displacement can be measured by attaching a gauge between the platform at the bottom and the loading head on the top. If manual recording is adopted, increase the axial load by increments at approximately 5% of the estimated strength and take readings in each step. Continue the process until failure.

To dismantle the test assembly, release both the axial and confining pressure simultaneously and slowly while someone holds the triaxial cell assembly to avoid falling. Remove the assembled triaxial cell carefully from the loading frame and lay it on a table. The failed specimen can then be removed from the cell. A light pressure may be needed to press the specimen out of the cell. Care must be taken to avoid damaging the membrane.

Make a sketch or take a photo of the failed specimen.

Repeat the above process to test another specimen at a different confining pressure, until all specimens are tested.

6. Calculations and reporting

Transfer all data to an excel sheet with time, load and axial displacement (if measured) listed sequentially in separate columns. Calculate the axial-confining stress difference from the axial load

$$(\sigma_1 - \sigma_3) = (P - P_o) / (\pi d^2 / 4) \qquad (A.3)$$

If axial displacement is recorded, convert it to axial strain by the following equation (unless strain is recorded)

$$\varepsilon = \delta / L \qquad (A.4)$$

where L is the specimen length and δ is the axial displacement. With above data, a $(\sigma_1 - \sigma_3) \sim \varepsilon$ curve as shown in Fig. 2.8 may be plotted. The highest point on each curve is used to calculate the corresponding triaxial strength.

If no axial displacement or strain is measured, the axial load at the failure point is used to determine the corresponding triaxial strength.

For tests at various confining pressures, a set of data $(\sigma_1 - \sigma_3)$ at the failure are now available. Plot them as Mohr's circles as shown in Fig. 3.7 and determine the values of (c_i, ϕ_i).

Estimate the angle Ψ of the failure plane of each specimen based on the observation of failure as shown in Fig. 5.4 and take an average. Determine the friction angle ϕ_i again from Eqn. 5.7b

$$\phi_i = 90° - 2\Psi \qquad (A.5)$$

Compare the friction angle values from the above two methods and discuss the results.

Subject Index

About the author: Daihua Steve Zou, a Professor of Mining Engineering at Dalhousie University, has more than 35 years of combined experience in industry, research and teaching in mining engineering. He was awarded an Honorary Professorship at Shandong University of Science and Technology in 2006, and was appointed as an Adjunct Professor at Beijing University of Science and Technology in 1999, Chinese Academy of Science, Chengdu Mountain Hazards Research Inst. in 2007, and Chengdu University of Technology in 2016.

Steve was born in a small mountain village at the edge of Chengdu Plain in Sichuan, China during the 1950's. From a young age, Steve was eager to learn and determined to pursue an education. In 1977 when higher education resumed after the *"Culture Revolution"*, he ranked in the top 1% in the nation-wide university entrance exams and was accepted to China University of Mining and Technology, where he graduated at the top of his class with a B.Sc. in Mining Engineering in 1982. In the same year, Steve competed in national postgraduate entrance exams and was awarded the opportunity to study in University of British Columbia with a full scholarship, where he earned a Ph.D. in mining rock mechanics in 1988. Dr. Zou conducted post-doctoral research at Laurentian University prior to joining Kidd Creek Mine as a Geomechanics Engineer in 1990. He then joined Technical University of Nova Scotia (now part of Dalhousie University) as an Assistant Professor in 1992 and was promoted to a full Professor in 2001. He has been teaching mining engineering and supervising Master's and Ph.D. students for 28 years. He served as the Mining Engineering Program Chair and was the founding Head of the Department of Civil and Resource Engineering until 2010.

Professor Zou has been active in research as well, engaged in many research projects and published over 80 papers in the referred journals and conference proceedings. His research includes rock mechanics, slope and underground stability, mine excavations, in-situ stress measurement, numerical modeling, non-destructive rock bolt monitoring and dry disposal of mine tailings. He lately collaborated with international researchers in geohazards research on debris flows and landslides.

Professor Zou is well recognized internationally in his field. He has been invited on many occasions by international institutions and governmental agencies to serve as an expert in evaluating academic programs, research and industrial projects, and publications. He has played pivotal roles in organizing several international conferences and is the founder of the International Journal of Geohazards and Environment.